Plant Life Histories re-examines patterns of relationships between plant life history traits in phylogenetic perspective. The re-examination recognizes that because evolution is a branching process, traits are not randomly distributed across taxa and that therefore analysis of trait correlations cannot treat species as independent data points.

Part 1 looks at the use of the phylogenetic perspective on trait correlation. Parts 2 to 4 examine traits from the reproductive phase, seed production and dispersal to recruitment and growth. The final section looks at interactions between plants and competitors, herbivores and microbial symbionts, recognizing that these interactions may have an ancient evolutionary history. Students and researchers of evolution, ecology and botany will find much of value here.

Plant Life Histories

Plant Life Histories

Ecology, Phylogeny and Evolution

Edited by

Jonathan Silvertown

Miguel Franco

and

John L. Harper

CAMBRIDGE
UNIVERSITY PRESS

PUBLISHED BY THE PRESS SYNDICATE OF THE UNIVERSITY OF CAMBRIDGE
The Pitt Building, Trumpington Street, Cambridge CB2 1RP, United Kingdom

CAMBRIDGE UNIVERSITY PRESS
The Edinburgh Building, Cambridge CB2 2RU, United Kingdom
40 West 20th Street, New York, NY 10011-4211, USA
10 Stamford Road, Oakleigh, Melbourne 3166, Australia

First published 1997

Printed in the United Kingdom at the University Press, Cambridge

Typeset in Times NR MT 9.5/13.25pt VN

A catalogue record for this book is available from the British Library

Library of Congress Cataloguing in Publication data
Plant life histories: ecology, phylogeny, and evolution / edited by
Jonathan Silvertown, Miguel Franco, and John L. Harper
     p.   cm.
Includes bibliographical references (p.   ) and index.
ISBN 0 521 57495 1 (pb)
1. Plants–Phylogeny.   2. Plants–Evolution.   3. Plant ecology.
I. Silvertown, Jonathan W.   II. Franco, Miguel (Franco Baqueiro)
III. Harper, John L.
QK989.P58   1997
581.3′8–dc21   96-51104   CIP

ISBN 0 521 57495 1 paperback

Contents

III Seeds

IV Recruitment and growth

V Interactions

Contributors

David Ackerly, *Department of Biological Sciences, Stanford University, Stanford, CA 94305, USA*

Spencer C. H. Barrett. *Department of Botany, University of Toronto, Toronto, Ontario, M5S 3B2, Canada*

Thomas Bataillon, *Department of Biology, McGill University, 1205 Avenue Docteur Penfield, Montreal, Quebec, H3A 1B1, Canada*

M. J. Crawley, *Department of Biology, Imperial College, Silwood Park, Ascot, SL5 7PY*

Mike Dodd, *Department of Biology, The Open University, Milton Keynes, MK7 6AA*

Michael Donaghue, *Harvard University Herbaria, 22 Divinity Avenue, Cambridge, MA 02138, USA*

A. H. Fitter, *Department of Biology, University of York, York, YO1 5YW*

Miguel Franco, *Instituto de Ecología, Universidad Nacional Autónoma de México, Apartado Postal 70-275, 04510 Coyoacán, D.F., Mexico*

Douglas J. Futuyma, *Department of Ecology and Evolution, State University of New York, Stony Brook, NY 11794-5245, USA*

M. J. W. Godt, *Department of Botany, University of Georgia, Athens, GA 30602, USA*

Deborah E. Goldberg, *Department of Biology, University of Michigan, Ann Arbor, MI 48109-1048, USA*

J. L. Hamrick, *Department of Botany, University of Georgia, Athens, GA 30602, USA*

Lawrence, D. Harder, *Department of Biological Sciences, University of Calgary, Calgary, Alberta, T2N 1N4, Canada*

John L. Harper, *Cae Groes, Glan-y-Coed Park, Dwygyfylchi, Conwy, North Wales, LL34 6TL*

P. H. Harvey, *Department of Zoology, University of Oxford, South Parks Road, Oxford, OX1 3PS*

R. J. J. Hendriks, *Department of Ecology, University of Nijmegen, Toernooiveld, NL-6525 ED Nijmegen, The Netherlands*

L. Klimeš, *Section of Plant Ecology, Academy of Sciences of the Czech Republic, Dukelská, 379 82 Třeboň, Czech Republic*

J. Klimešová, *Section of Plant Ecology, Academy of Sciences of the Czech Republic, Dukelská, 379 82 Trěboň, Czech Republic*

Michelle Leishman, *School of Biological Sciences, Macquarie University, NSW 2109, Australia*

Janice Lord, *Botany Department, University of Otago, Dunedin, New Zealand*

Charles Mitter, *Department of Entomology, University of Maryland, College Park, MD 20742, USA*

Martin T. Morgan, *Department of Biology, McGill University, 1205 Avenue Docteur Penfield, Montreal, Quebec, H3A 1B1, Canada*

B. Moyersoen, *Department of Biology, University of York, York, YO1 5YW*

A. Purvis, *Department of Biology, Imperial College, Silwood Park, Ascot, SL5 7PY*

Mark Rees, *Department of Biology, Imperial College, Silwood Park, Ascot, SL5 7PY*

Daniel J. Schoen, *Department of Biology, McGill University, 1205 Avenue Docteur Penfield, Montreal, Quebec, H3A 1B1, Canada*

Richard M. Sibly, *School of Animal and Microbial Sciences, Whiteknights, PO Box 228, Reading, RG6 6AJ*

Jonathan Silvertown, *Department of Biology, The Open University, Milton Keynes, MK7 6AA*

J. M. van Groenendael, *Department of Ecology, University of Nijmegen, Toernooiveld, NL-6525 ED Nijmegen, The Netherlands*

D. Lawrence Venable, *Department of Ecology and Evolutionary Biology, University of Arizona, Tucson, AZ 85721, USA*

Mark Westoby, *School of Biological Sciences, Macquarie University, NSW 2109, Australia*

Anne C. Worley, *Department of Botany, University of Toronto, Toronto, Ontario, M5S 3B2, Canada*

Preface

The great Darwinian truth that underlies our attempts to discover rhyme and reason in the diversity of life on Earth is that natural selection has shaped the form and behaviour of organisms. The search for the evolutionary pathways that lead to the present diversity of life, the study of phylogeny, was among the most powerful forces in the development of biological science in the latter part of the nineteenth and first half of the twentieth century. It provided a fascinating intellectual exercise to draw out putative evolutionary sequences and gave the excuse for quite violent conflicts of interpretation. Comparative morphology was the main (and often the only) source of data for such phylogenetic speculation. It acknowledged that some features, 'conservative characters', were more resistant than others to evolutionary pressures and so were more reliable for tracing lineages. To establish phylogenies it became vitally important to identify these 'conservative' characters and to distinguish them from features that responded more quickly to selective pressures and therefore indicated only recent ancestry.

Two particularly important developments stemmed from the historic passion for phylogeny. First, and of great significance for many chapters in this book, was the perceived need for systematics and taxonomy to reflect evolutionary lineages and to be more than efficient systems for classification and pigeon-holing. A second development from the post-Darwinian passion for phylogenetic speculation was a deep distrust of what had sometimes become 'the gin-palaces and brothels of unbridled hypothecation'. Disillusionment with some of the more extreme phylogenetic 'just-so stories' forced a quite dramatic shift in interest from macroevolutionary speculation to microevolutionary empiricism. Evolution in action became the focus for the study of biological diversity. The science of ecological genetics, pioneered for example by E. B. Ford among zoologists and Turesson among botanists, emphasized the role of present environmental forces (both biotic and abiotic) in natural selection. Studies of melanism in moths and metal tolerance in plants were just two examples among many that emphasized the power and speed with which natural selection operated. The process of evolution had become part of the present ecological scene. It has been tempting to see the ecological match between organisms and their present environments as the result of present evolutionary processes, and to forget their phylogeny.

This book was born out of the conviction that the time is ripe to re-examine in phylogenetic perspective the perceived patterns of relationship between different plant life history traits, and between those traits and the presumed selective pressures that shaped them. This re-examination takes two forms. In the first it simply recognizes that because evolution is a branching process, traits are not randomly distributed across taxa and that therefore analyses of trait correlations cannot treat species as independent data points. One solution to this problem is essentially (though not literally) to factor-out phylogeny so that trait correlations that arise from convergent evolution can be distinguished from those due to common descent.

The second form of re-examination is almost the reverse of the first and, instead of attempting to factor-out phylogenetic relationships between species it uses these relationships to reconstruct the evolutionary pathways of traits. Phylogeny allows us to infer whether a particular change occurred once or a number of times, whether change has been in one direction or another, whether change in one character has been associated with change in another, whether associated changes have occurred in a particular sequence, and whether particular changes are associated with shifts in diversification rates.

The volume is organized in five parts. The first section is chiefly intended to demonstrate the uses and hazards of taking a phylogenetic perspective on trait–trait and trait–habitat correlations. Parts two, three and four examine a range of traits from the reproductive phase of the life cycle through seed production and dispersal to recruitment and growth. In each of these parts, the chapters variously emphasize the two forms of phylogenetic perspective, trait–trait or trait–habitat correlations, and relevant life history theory. The fifth and final part recognizes that important features of plant life history involve interactions with competitors, herbivores and microbial symbionts and that these interactions have an evolutionary history – sometimes an ancient one.

Phylogenetic perspectives

In an introductory chapter Silvertown and Dodd ask what they see as the fundamental questions in any comparative study of plant life histories: 'Which traits are correlated with each other?' and 'Are these correlations the result of common descent or have they arisen repeatedly as a result of convergent evolution?'. They distinguish between comparisons that treat species as independent data points (the tips of the phylogenetic tree, TIPs) and those that take account of the phylogeny and the lack of independence that common ancestry may confer on samples (phylogenetically independent contrasts, PICs). This theme appears repeatedly in other chapters in this book and Silvertown and Dodd show, with some worked examples, how phylogenetic information can be used to correct comparisons for phylogenetic independence. They show that secondary chemistry and life form (woody *vs* herbaceous), two traits that are conserved in angiosperm phylogeny, are significantly correlated due to convergent evolution, providing important support for theories of anti-herbivore defence. Lest ecologists rely too much on specific evolutionary trees Donoghue and Ackerley caution that 'It is highly likely that, in fact, virtually every phylogenetic tree found in the literature is wrong in one way or another, and they describe the use of sensitivity analyses to explore the robustness of phylogenetic conclusions.

A major part of the eco-evolutionary game has been explored by demonstrating what it is about particular species that enables them to live where they do and what it is about them that makes them seem the best (or the best compromises) in the best of all possible worlds. Extraordinarily little effort has been put into ecological and evolutionary explanations of why almost all species are always absent from almost everywhere! One explanation is presumably that evolutionary specialization progressively narrows the range of habitats and environments that provide the conditions needed for survival and growth. However, another reason for the absence of most species from most habitats is that geographic isolation has prevented colonists from ever reaching them.

Species of plant and animal that are new invaders to a territory, especially to an island, must of necessity have acquired the necessary properties elsewhere. They cannot become subject to natural selection in their new habitat until after they have proved that they can survive in it. Comparisons between alien and native floras and faunas are therefore a particularly rich source of material for studying the nature and origins of the traits that govern the ecology of species. Crawley, Harvey and Purvis use this approach for an analysis of the British flora for which there is detailed information about the time of alien invasions. Using PIC they find traits such as the mating system and method of seed dispersal to be apparently irrelevant to successful invasion and that successful invaders are characterized by being big (large seeds, large plants) and having protracted seed dormancy. Their analysis also suggests that the flora of Britain is unsaturated because new invasions have not depended upon the ousting of previous inhabitants.

Reproductive traits

A wide variety of mating systems is found among flowering plants, often within the same family or genus, and so even before the advent of molecular phylogenies it had become clear that shifts between selfing and outcrossing must have occurred frequently within lineages. Molecular phylogenies now make it possible to unravel the story of mating system evolution in much greater detail. Barrett, Harder and Worley compared species within the family Polemoniaceae to determine the degree of independence of phylogenetic and ecological correlates of self- or cross- fertilization. Using a molecular (*matK*) phylogeny they show that in the temperate herbaceous clade of Polemoniaceae the annual life cycle evolved at least seven times, and reverted to the perennial habit on three occasions. Selfers evolved from outbreeders at least 14 times but there is no evidence of the reverse.

Schoen, Morgan and Bataillon examine the clues that both floral ecology and molecular genetics offer to answer the puzzling question of why self-pollination has evolved repeatedly within different lineages. Their analysis of various theories shows that a century of study has left us still in the dark about this central problem in plant reproductive biology (as in our understanding of the significance of sex itself). While the first two chapters in this part deal with mating system as an evolving trait, the third chapter by Hamrick and Godt analyses the effect of this trait, amongst others, on genetic diversity and genetic structure.

Seeds

Seeds for the plant scientist are (like eggs for the zoologist) convenient starting points for the analysis of life histories. There is abundant evidence from simple controlled experiments that large seeds confer a clear advantage in intraspecific competition, and that individuals whose seedlings are smaller or emerge later are commonly disadvantaged in a population of the same (or a closely related) species. Seed mass is part of a classic example of evolutionary compromise in which any increase in individual seed mass is thought necessarily to involve a compensating decrease in seed number. Increased seed mass almost inevitably makes dispersal more costly but by providing the embryo with greater reserves increase the chance of establishment in environments where resources are hard to gather. Delayed germination implies that time spent dormant might, in theory, have been employed in growth and reproduction and the disadvantage is usually assumed to be offset by an increased probability of survival as a result of the delay. There are extensive data sets from the British flora which Rees analyses using PICs to explore the correlations between seed mass, dormancy, life span and clonality, finding general support for theoretical expectations.

The chapter by Westoby, Leishman and Lord is broadly concerned with detecting correlations among the same series of traits but provides a contrast with Rees' paper because greater weight is given to the informative value of TIPs analyses. PIC only detects evolutionary change and is hence insensitive to the consequences of stabilizing selection. Westoby *et al.* make the important point that variation in seed size within species is typically very narrow, implying that in each species seed size (mass) is tightly canalized and presumably subject to strong stabilizing selection. In contrast the mass of individual seeds differs quite dramatically from species to species within the same communities. It looks as though natural selection may act powerfully to favour uniformity within species but ecological forces act to favour seed niche diversity within communities.

A crucial question is when and whether any property (form or behaviour) of an organism, population or species can usefully be analysed in isolation from others. Perhaps the most interesting 'traits' are trade-offs in which the balance between two or more attributes is more interesting (both for phylogeny and ecology) than each attribute itself. The relationship between the size and number of seeds may, for example, be a far more informative 'trait' than either size or number alone. Venable discusses models that can be used to help understand how and when different components of the reproductive process constrain and selectively impact each other.

Recruitment and growth

Early land plants grew clonally and clonality is clearly an ancestral trait in the plant kingdom. We need therefore to explain how and when it has become lost from some groups and what the ecological consequences are. Van

Groenendael *et al.* analyse both the phylogeny (at the level of families and higher systematic categories) and the present ecology of clonality using a unique dataset of 2000 plants from Central Europe. One important finding is that clonality is overrepresented among plants of wet habitats, but that this is essentially due to the preponderance of monocots in such habitats.

The whole pattern of a life history can be regarded as a trait, selected in evolution and constraining the ecology of species. But, of course, most individuals never experience their life history: most die young. However, among those that survive, the pattern of adult mortality differs quite markedly between species and various authors have recognized, among animals, a continuum of life histories ranging from organisms with 'fast' (high adult mortality, fast development, high fecundity and short life cycles) to 'slow' life histories (low adult mortality, slow development, low fecundity and long life cycles). Franco and Silvertown ask whether such a continuum can be detected in the life histories of higher plants and they conclude that the modular growth of plants (which enables them, at least in theory, continuously to increase their size and fecundity as they age) means that some trait correlations found in unitary organisms are not seen in plants.

We may recognize individual species as having characteristic life histories that we can classify or order. But in reality nature is made up of environments that are heterogeneous, patchy mosaics and in these genetically identical individuals act out their life histories in quite different fashions. The 'characteristic' life history of a species is defined by a variance as well as a mean. Sibly reviews the body of theory that has developed to model the evolution of life histories in environments that are heterogeneous in space and also in time. Once again, there are peculiarities about higher plants, such as their plasticity and the potential increase in fecundity with age, that make classic models based on unitary animals inappropriate.

Interactions

Ecologists see competition, predation and parasitism as powerful forces in present population and community dynamics that determine the present (proximal) significance of traits in both form and behaviour. It would be unreasonable to disregard their potential role in phylogeny. It is to be expected that past biotic interactions (ultimate factors) will have left their mark in present (proximal) performance as limits and constraints on the ecological versatility of species. A classic paper by Ehrlich & Raven (1964) on the evolutionary consequences of interactions between insects and their food plants was seminal in the development of the theory of coevolution. Insect herbivores are conservative in their diet often to the extent of strict mono-phagy. Indeed, biological control programmes that use insects to control weeds depend on this monophagy for their success. Such dietary specializ-ation is often deeply rooted in the history of the clades of both the insect herbivores and in their plant prey where it is often reflected in biochemically versatile modifications of particular classes of alkaloids, glucosinolates, ter-penes etc. Futuyma and Mitter review the state of knowledge about the evolution of dietary specialization among insects and point out, for example, that some cladistically basal groups that feed on conifers and other gymnos-perms have probably been associated with their special hosts since before the Cretaceous. Where there is evidence that specialists moved to new host plants and lineages have consequently branched, movement has usually been from one food plant to another that is closely related. There are few better demon-strations of phylogenetic constraint acting upon what selection can accom-plish.

The associations between microbial symbionts and plant roots are even more ancient than the majority of plant–herbivore associations. Fitter and Moyersoen discuss the phylogeny of root symbioses and point out that endo-mycorrhizal associations are primitive but the non-mycorrhizal habit is a derived, specialized condition that appears to have evolved repeatedly. Ectomycorrhizal associations have evolved at least twice. By contrast, the ability to form N-fixing root symbioses appears to be restricted to a single clade but within this clade the ability may be gained or lost.

Haldane (1932) wrote 'The fitness of plants in the Darwinian sense must be tested with the plants grown in competition' and, as Sakai (1961) pointed out, 'This stimulating statement implies the existence in plants of heritable dif-ferences in competitive ability.' Sakai went on to compare the performance of varieties of rice in pure stands and in all combination of varieties in pairs. From this simple experimental design he concluded that competitive ability is a genetic character controlled by polygenes with low heritability. Since that time the study (even the definition) of plant competition has developed into a minefield of problems – some semantic and some that pose daunting prob-lems in experimental design and analysis. Competitive ability might be re-garded as a trait in itself; it clearly also encapsulates a whole spectrum of individual traits that interact to determine success or failure in a struggle with neighbours. There is certainly no way in which we can hope to identify traits associated with competitive effect and competitive response or to partition 'competitive ability' into proximate (ecological) and ultimate (phylogenetic) elements until we have some agreement on what it is that we are measuring and comparing. Goldberg reviews many of these problems with a succinct survey of a vast literature and suggests simple (though very large) experimental designs that might provide a solution.

A phylogenetic perspective reminds us that there is a unity of common origins underlying the diversity of plant life history. This volume is intended to uncover some of that unity, but in the attempt it is inevitable that the holes in the fabric of our knowledge become apparent too. Lest the holes seem larger than the cloth, we should pinpoint some of the more general trait correlations revealed by the use of the comparative approach in this volume: longevity is a trait of central importance that is negatively correlated with reproductive allocation (Silvertown & Dodd) and seed dormancy (Rees), and positively correlated with outcrossing (Barrett *et al*), genetic diversity (Hamrick & Godt), age at first reproduction (Franco & Silvertown), and seed mass (Rees). Seed mass is another very important trait that is correlated negatively with dormancy (Rees), specific leaf area and relative growth rate (Westoby *et al.*) and positively among herbs with the presence of VA mycorrhizas (Fitter & Moyersoen). Life form (herbaceous/woody) is a third trait of general importance, being correlated with leaf chemistry (Silvertown & Dodd), mating system (Barrett *et al.*) and ecotomycorrhizas (Fitter & Moyersoen). This is an incomplete list and not all the correlations are simple ones (for example the negative relationship between seed mass and dormancy occurs only in species with specialized dispersal mechanisms), so one of the most important questions we have to ask is 'How general are these patterns?'. Some, like the correlations with longevity, may be quite general, others like those involving symbionts may be more clade- or habitat-specific. Although it is still too early to answer the question, this volume demonstrates that now at least we have the means to address it.

John L. Harper
Jonathan Silvertown
Miguel Franco

References

Ehrlich, P. & Raven, P. H. (1964). Butterflies and plants: a study in coevolution. *Evolution*, **18**, 586–608.

Haldane, J. B. S. (1932). *The causes of evolution*. London: Longman.

Sakai, K. (1961). Competitive ability in plants. In *Mechanisms in biological competition*. Symposia of the Society for Experimental Biology, **15**, 245–263.

I · Phylogenetic perspectives

1 • Comparing plants and connecting traits

Jonathan Silvertown and Mike Dodd

1 Introduction

Plants arguably display a greater range of life history variation than is found in any other kingdom and this diversity provides a wealth of raw material for comparative studies on evolution and ecology. In this chapter we will first make some general and methodological points about using the comparative method to seek trait correlations in plants and then describe two new comparative analyses of our own by way of illustration. The closing remarks will consider the relative merits of phylogenetic *vs.* non-phylogenetic methods of comparative analysis.

The two fundamental questions for any comparative study are:

1. Which traits are correlated with one another?
2. Are trait correlations the result of common descent or of convergent evolution?

The first of these questions is routinely asked by anyone interested in plant life history, whether they are primarily interested in an ecological or an evolutionary interpretation of the answer, but the second question is asked by plant ecologists much less frequently than it ought to be. Why, many plant ecologists seem to ask, do we need to worry about phylogeny at all if we are not asking an evolutionary question? More subtly, Westoby *et al.* (1995 and this volume) have argued that phylogenetically based comparative methods are conservative in their interpretation of the evidence for adaptation and that correlations based on species as independent data points give the *ecological* independence of species from one another their due weight. Harvey (1996, Harvey *et al.* 1995) has recently answered this question quite comprehensively, so the answer given here will be as brief as possible.

A typical ecological question is: Do the species found in one type of habitat have larger seeds than the species found in another type of habitat? Several studies have addressed this sort of question and, using each species in their samples as an independent data point, they have concluded, for example, that species living in dry habitats have larger seeds than those in wetter habitats (Baker 1972), and that light-demanding pioneer tree species have smaller seeds than species of mature tropical forest (Foster & Janson 1985). If one did not want to interpret these correlations causally, then there would be no problem with this approach, but of course one *always* wants to explain *why* mean seed size differs between habitats and it is this that produces the problem. Since all species have a common ancestor at some point in their evolutionary history, there will always be some degree of ancestral similarity between the species sharing a habitat, and common ancestry, not adaptation, may be responsible for similar seed size. Species sharing a habitat may also have similar seed sizes as an accidental consequence of an association between seed size and some other trait. To put it in statistical terms, common descent causes traits to be confounded with one another when species are the data points. If two species share similar traits due to common descent, then treating them as independent data points when testing for trait correlations is the equivalent of pseudoreplication in a designed experiment (Rees 1995).

There seems to be a common misconception that the argument just summarised applies only to causal hypotheses that invoke natural selection (e.g. shade selects for large seed size), and that it does not invalidate tests of hypotheses invoking ecological sorting (e.g. only large-seeded species can colonise shaded habitats) as a cause of trait/habitat correlations (Westoby *et al.* 1995). However, if common descent confounds traits, any test for trait correlation that ignores this will be jeopardised, regardless of the causal hypothesis being tested.

The problem, then, is how to deal with the confounding effects of common descent. The best solutions to this problem involve using a phylogeny for the taxa being compared. Take a phylogenetic tree for a group of four species A–D in which we want to test whether the value of two traits are correlated (Figure 1.1). Felsenstein (1985) pointed out that differences between species A and B will be independent of differences between C and D because in each case those differences arose after the common ancestor of all four species lying at the node marked 'g'. We can therefore treat those differences, technically called contrasts, as phylogenetically independent of one another. Furthermore, if we use an evolutionary model to estimate the value of the traits at nodes 'e' and 'f', then the contrast between these nodes is also phylogenetically independent. By working down an entire phylogenetic tree in this way, a phylogeny for n species can provide $n-1$ contrasts for hypothesis testing. By testing for correlation (for quantitative traits) or association (for qualitative

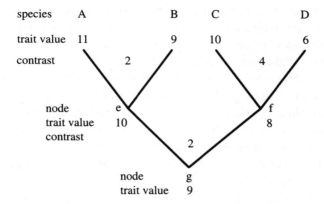

Figure 1.1. A phylogenetic tree for four species (A–D). Contrasts are calculated as the difference between the trait values of the two species that branch from a common node. Trait values may be assigned to internal nodes of the tree by assuming an evolutionary model of character evolution. Under the Brownian motion model, a node's trait value is the mean of the values for the two species (or nodes) that branch from it. In the example shown, all contrasts are positive, but in reality some traits will show negative contrasts on the same tree. Also, in this example it has been assumed that branch lengths are equal, though the method can use branch length information if this is available.

traits) between trait contrasts instead of between trait values, the effects of common descent can be removed.

If there is no phylogeny available, then one solution to the problem is to confine comparisons to congeneric or confamilial pairs of species because we can usually be pretty confident that these share a more recent common ancestor than species in other genera or other families (the only doubt can be over whether taxonomic classification accurately reflects phylogeny). Salisbury (1942) used this technique on congeners to look for associations between seed size and habitat many years ago and, more recently, Kubitzki & Ziburski (1994) used intrageneric comparisons to show an association between seed dispersal type and habitat in tropical species for which no phylogeny was available.

2 Examples

2.1 Reproductive allocation

Our first example looks at patterns of reproductive allocation (RA) and asks: Does the proportion of dry weight devoted to reproductive structures vary with plant life history and does it vary with the successional stage of habitats? These were fashionable questions 20 years ago, but the steam rather went out of the subject when people began to compare results for different species and from different studies and realized that there were a host of methodological questions, such as whether root biomass had been measured or not, that made it difficult to draw general conclusions. We do no more than mention the methodological problems here because they are secondary to our purpose and they have been comprehensively reviewed by Willson (1983). In a quandary, she noted that exceptions to the expected patterns were legion: 'We do not know if these are unusual populations of the species or if these plants are true exceptions to the "rule".'

Suffice it to say, many of the difficulties can be overcome if comparisons are designed to control for the phylogeny of the species and for the authorship of the study. Surprisingly, neither of these controls has been systematically applied in a comparative study of plant reproductive allocation before. Phylogenetic control can ensure that the major differences between taxa (for example, in the size of the infructescence and other ancillary reproductive structures) do not overwhelm differences that correlate with life history. By confining comparisons to those possible within the results of individual studies, we can also eliminate at least some of the methodological differences between authors.

One of the largest and most frequently cited studies of reproductive allocation was by Abrahamson (1979) who measured RA in 50 herb species collected from old fields and woods in Pennsylvania. He reported that species of old fields (i.e. early succession) had a higher RA than did woodland herbs (later in succession), and also that old field annuals had a higher RA than old field perennials, but his analysis treated each species as an independent data point. This comparative method is referred to as TIPS (because all the tips of the phylogenetic tree are compared), in contradistinction to PIC (phylogenetically independent contrasts).

We have tested the two patterns reported by Abrahamson using PIC on his data, supplemented by data from other sources. Thirteen of the 50 species in Abrahamson's study belong to the family Asteraceae, though only one of these occurred in woods. The phylogeny of these species (Figure 1.2), based on the molecular phylogeny of Jansen et al. (1990), illustrates three contrasts that can be used to test the patterns reported by Abrahamson. Neither the RA contrast between field and woodland species of Hieracium ($t_{18} = 0.745$, $P > 0.05$) nor the contrast between the annual Erigeron annuus and the mean RA of the two most closely related perennials ($t = 0.94$, $P > 0.05$) is signifi-

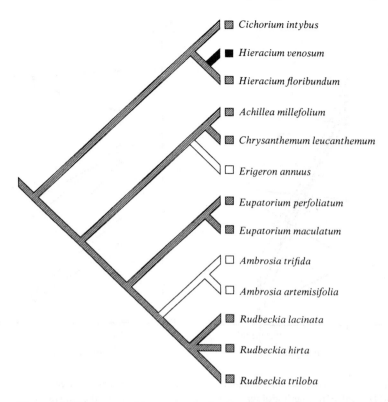

Figure 1.2. Phylogeny of Asteraceae in the Abrahamson (1979) dataset. Annuals shown in white, old field perennials in grey and the single woodland perennial in black. Two annual/perennial contrasts are present, and one field/woodland habitat contrast. The tree was drawn by combining the phylogenetic relationships among tribes given by Jansen *et al.* (1990) with the tribal membership of species given by Mabberly (1987).

cant. The contrast between the mean RA's of annual *Ambrosia* spp. (8.5%) and perennial *Rudbeckia* spp. (16%) is significant, but in the wrong direction for the hypothesis ($t = 6.39$, $P < 0.001$).

One other contrast between annual and perennial species in the same family (Primulaceae) was available in Abrahamson's dataset: the contrast between the annual *Anagallis arvensis* (RA = 38 ± 12.4) and the perennial *Lysimachia ciliata* (RA = 3.0 ± 1.5) was highly significant ($t_{18} = 8.86$, $P < 0.001$) in the expected direction. We also looked for annual/perennial contrasts in other studies that could be used to test this pattern and found 10 further pairs distributed among eight families. Analysed as a group (including

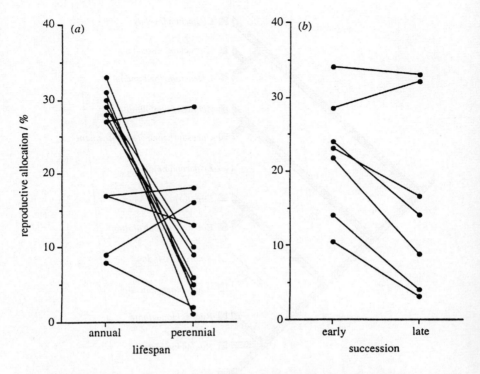

Figure 1.3. (*a*) RA contrasts for 13 annual/perennial pairs of species (some points overlap). To control for methodological differences between studies, the contrasts shown are confined to species where the same authors had measured RA in both the annual and the perennial. (*b*) RA contrasts between seven pairs of early and late successional perennial species in the Asteraceae. Data are from various sources, listed in Appendix 1 of Hancock & Pritts (1987) and include one contrast from Abrahamson (1979).

the Abrahamson contrasts), these data strongly support the hypothesis that annuals have a higher RA than perennials (Figure 1.3*a*) (Wilcoxon matched pairs test, $n = 13$, $t = 8$, $P < 0.009$). The data of Wilson & Thompson (1989) on the RA of annual and perennial British grasses were not included in the analysis, but also strongly support this result.

A total of five confamilial contrasts was available from Abrahamson's data to test his successional pattern. Two of these contrasts were significant in the predicted direction and three were not, but overall the null hypothesis of no association between habitat and RA could not be rejected (Wilcoxon matched pairs test, $n = 5$, $t = 5$, $P = 0.50$).

Hancock & Pritts (1987) reviewed the literature relevant to Abrahamson's two hypotheses and tabulated the RA of 30 perennial species belonging to the Asteraceae, 11 of which are found in early successional habitats and the remainder in later succession. Confining comparisons to those within individual studies as before, this dataset includes seven phylogenetically independent contrasts between early and late successional species (Figure 1.3b). If phylogeny and authorship are ignored, there is no significant difference in RA between the two groups, but when data points are matched, the difference is significant in the direction predicted by Abrahamson's hypothesis (Wilcoxon matched pairs test $n = 7$, $t = 2$, $P = 0.043$).

The phylogenetically-controlled tests of Abrahamson's two hypotheses demonstrate how independent contrasts can be used to clarify what were two very messy relationships based on pseudoreplicated samples (i.e. lacking in phylogenetic control) and with many apparent exceptions. Exceptions remain of course, but even these may prove to be more informative now that the patterns have been soundly established.

2.2 Secondary chemistry and life form

Our second example is altogether a much more ambitious one. Rapid progress is now being made in constructing a molecular phylogeny for the families of flowering plants, and this provides the exciting opportunity to look at large-scale life history patterns across the angiosperms as a whole. In 1993 Chase and 41 coauthors published phylogenetic hypotheses for 265 families of angiosperms based on DNA sequences of the *rbc*L gene in nearly 500 plant species (Chase *et al.* 1993). Though subsequent work has led to a revision of some of the phylogenetic relationships originally proposed by Chase *et al.* (Rice *et al.* 1995), it still forms an invaluable basis for exploring trait variation at the family level and above.

We have used the independent contrasts method and the Chase phylogeny to test the hypothesis advanced by Feeny (1976) that the types of defensive secondary compounds deployed by plants differ according to plants' 'apparency' to herbivores. Feeny (1976) suggested that plants that are more 'apparent' to herbivores should invest in relatively high tissue concentrations of digestibility-reducing compounds such as tannins, while plant species that are less apparent should invest in more toxic compounds such as alkaloids that are poisonous at relatively low concentration. Although this theory is in many ecological textbooks and has been cited over 635 times in the primary literature, there has hitherto been no comprehensive test of the hypothesis.

Our test utilizes data on the phylogeny and secondary chemistry of over 165 angiosperm families.

An objection that has been raised against the apparency hypothesis is that it is difficult to know what particular features of a plant make it 'apparent' from a herbivore's point of view. For the purposes of the present test we follow Feeny in assuming that woody plants will be more apparent than herbs because of the gross difference in size between the two life forms. If this assumption is incorrect, it is unlikely to bias the test in favour of a positive result. Another objection raised is that the association between the tree life form and the presence of tannins, which Feeny noted, could be the product of common descent (Mole 1993). Life form and most kinds of secondary compounds, including the tannins and alkaloids, are phylogenetically conservative in their distribution, so comparative tests that seek correlations among these variables must establish the phylogenetic independence of any relationship that is claimed.

We used PIC as implemented in the CAIC computer package (Purvis & Rambaut 1995) to analyse our dataset. It is important that the phylogeny used in any such analysis has been derived completely independently of the traits (life form, tannins, alkaloids) under investigation so the Chase *et al.* molecular phylogeny was ideal for this purpose. As generally recommended for comparative analyses (Eggleton & Vane-Wright 1994), the published example of one of the most parsimonious trees (from Search II) was used in preference to the consensus tree. Fifteen polyphyletic families were omitted from the analysis. CAIC permits branch lengths to be treated in a variety of ways; we chose the option that sets them all equal because even though Chase *et al.* (1993) give branch lengths, it is known that the rate of evolution in the *rbc*L gene varies considerably between different angiosperm lineages (e.g. Bousquet *et al.* 1992a, Frascaria *et al.* 1993).

All data were analysed at the family level or above. We used two compilations of the taxonomic distribution of plant secondary compounds as sources of chemical data: Mole's (1993) tabulation of the proportion of tested species that contained foliar tannins in 224 families; and Levin's (1976) similar tabulation of the proportion of tested species containing alkaloids in 110 families. The tannin data were based on 2227 species and the alkaloid data on 11 299 species. Tropical as well as temperate families were well represented in both compilations. It has been claimed that alkaloids are more common in tropical families than in temperate ones (Levin 1976), but a test of this hypothesis using PIC on Levin's data did not support this pattern. Family synonymies were checked in Mabberly (1987) which was also the source of life-form data for each family. Families were scored as 'herbs present' or 'herbs absent', with the latter category including cases where Mabberly described herbs as 'few' or 'rare'.

In the three tests of the apparency hypothesis we asked:

1. Is the presence of the herbaceous growth form in a family negatively associated with the proportion of species in the family possessing foliar tannins? The answer was 'yes' (n = 36, $t = -2.836$, $P = 0.008$, 2-tailed test). It is worth noting that one of the contrasts that went in the 'wrong' direction in this test, because tannins were at low frequency in the woody taxa (Morningaceae + Caricaceae (woody) vs Brassicaceae and other herb families), fell within the clade of families that contain mustard oil glucosides (Rodman et $al.$ 1993). Mustard oils were the toxins Feeny (1976) originally had in mind as defensive alternatives to tannins.

2. Is the presence of the herbaceous growth form in a family positively associated with the proportion of species in the family possessing foliar alkaloids? A positive association was found, but this was not significant ($n = 21$, $t = 1.566$, $P = 0.133$, 2-tailed test). This result was heavily dependent on one contrast (Sterculiaceae vs Malvaceae) in the 'wrong' direction for the hypothesis. On average 24% of species in non-herb families contained alkaloids compared with 41% in herb families, but 40% of species tested in the Sterculiaceae (no herbs) and only 16% of species tested in the Malvaceae contained alkaloids. Removing this contrast gives $P = 0.053$ for a 2-tailed test.

3. Is the proportion of species in a family possessing foliar tannins negatively correlated with the proportion possessing alkaloids? The answer was 'yes' (Figure 1.4).

Sample sizes in these tests were smaller than the number of data points for individual variables because in tests 1 and 2 involving the categorical variable herbs present/absent, only contrasting families or branches were compared and in test 3 sample size was determined by the intersection of the alkaloid and tannin datasets.

Tests 1 and 3 strongly support the apparency hypothesis, test 2 offers weak (or no) support. Feeny's original observation in 1976, that foliar tannins are associated with the tree life form and are under represented among herbs, is robust and appears to be a genuine evolutionary pattern, not an artefact of common descent. The comparison of tannin distribution with alkaloid distribution (Figure 1.4) strongly supported Feeny's suggestion that these two different kinds of chemical defence are negatively associated. Although these

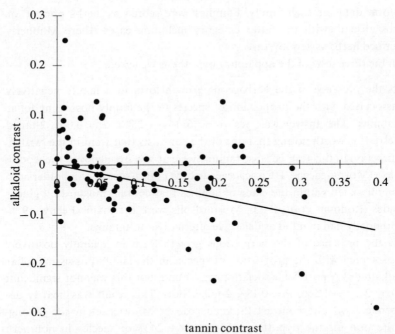

Figure 1.4. Relationship between alkaloid and tannin contrasts. Regression through the origin: $F_{1.77} = -21.22$, $P < 0.0001$.

results support Feeny's hypothesis, they do not refute other theories which make similar predictions (e.g. Coley *et al.* 1985).

A recent review of the theory and data on chemical defences concluded that apparency theory and many other hypotheses have all failed to provide a general explanation of why plants deploy particular kinds of chemical defence (Berenbaum 1995). Our test of the apparency hypothesis draws upon the widest possible sample of plants distributed right across the angiosperms, and its positive result is therefore cause for optimism that a general theory of plant chemical defences is indeed attainable. Broad, comparative phylogenetic tests of other theories are now called for.

3 Discussion

Phylogenetically independent contrasts has become the method of choice in comparative studies and the examples given demonstrate its utility in answering questions about life history evolution in plants. However, there is resistance to the adoption of PIC among some plant ecologists and the problems of applying phylogenetically correct methods do deserve serious attention. Modern comparative methods demand phylogenetic information that is at present still lacking below the family level for most plants. However, we should not let this blind us to the real problems of using TIPS. Computer simulation studies have demonstrated that TIPS (16%) has a much higher Type I error rate (rejecting a true null hypothesis) than PIC (5%) (Martins & Garland 1991). When a phylogeny is fully resolved the Type II error rate (accepting a false null hypothesis) is also higher for TIPS than for PIC, and PIC therefore has the greater statistical power (defined as 1 − Type II error rate). Purvis et al. (1994) applied PIC to simulated phylogenies varying only in the number of polytomies (unresolved relationships among branches). They found that the Type I error rate of PIC was quite robust to poor phylogenetic resolution, though the Type II error rate (and therefore statistical power) was badly affected.

Harvey & Pagel (1991) and others have suggested that a pragmatic solution to the absence of a phylogeny is to use taxonomic relationships as a surrogate, as a number of comparative plant studies have done (Kelly & Purvis 1993, Kelly 1995, Kelly & Beerling 1995; Kelly & Woodward 1995). Miles & Dunham (1993) caution that taxonomic trees may not be congruent with phylogenetic ones and should not be treated as surrogates for them. However their argument is weakened by the growing number of instances (although with notable exceptions) where molecular phylogenies are found to agree in large part with standard taxonomic treatments (e.g. Bousquet et al. 1992b, Bremer & Struwe 1992; Patterson et al. 1993; Bruneau et al. 1995).

More important than the congruence of taxonomy with phylogeny is the general point that phylogenies, as well as taxonomic surrogates for them, are only hypotheses and that many (sometimes many thousand) alternative interpretations of phylogenetic relationships are often equally parsimonious. Comparative analyses that use phylogenies should therefore test the sensitivity of their conclusions to the use of alternative phylogenetic hypotheses (Donoghue & Ackerly, this volume). At the moment this is a very arduous thing to do for large phylogenies, which is why we were unable to follow our own advice in the analysis of alkaloids, tannins and apparency, though improvements in software should soon make sensitivity analysis of this kind a practicality.

A concern often voiced by ecologists encountering PIC for the first time is that it is 'very wasteful of data' because there are many fewer contrasts than species in most datasets when binary variables such as annual/perennial are being analysed (the Abrahamson dataset discussed above is a typical example). The answer, seldom seen as helpful, is the point already made in the Introduction: that the degrees of freedom in TIPS analyses are inflated by pseudoreplication. A more helpful answer to the problem is the advice that datasets should be assembled with trait contrasts in mind so as to minimize the degree of redundancy, though of course one can only do this if the binary traits are known for a large number of species in advance; fortunately this is often the case.

Gittleman & Luh (1992) wisely suggest that before PIC is used in a comparative study, summary statistics on the data (e.g. nested ANOVA) should be used to determine whether phylogenetic effects are present for the traits of interest. If the influence of common descent is low (i.e. a high proportion of trait variance is found among species), then in the absence of a phylogeny for the group TIPS (or the equivalent at some higher taxonomic level) might be a reliable comparative method to use (e.g. Peat & Fitter 1994).

The problem of Type I error in TIPS, and the fact that most of the ecological patterns observed to date have been identified using this method, could be said to leave comparative plant ecology in the same state as a novel by Iris Murdoch: there are too many characters with too many unresolved relationships between them. The resolution of plant phylogeny promises also to resolve our picture of plant life history in all its aspects.

We thank the Open University Research Committee for financial support, and Paul Harvey and Kevin McConway for comments on the manuscript.

References

Abrahamson, W. G. (1979). Patterns of resource allocation in wildflower populations of fields and woods. *American Journal of Botany* **66**, 71–79.
Baker, H. G. (1972). Seed weight in relation to environmental conditions in California. *Ecology* **53**, 997–1010.
Berenbaum, M. R. (1995). The chemistry of defense – theory and practice. *Proceedings of the National Academy of Sciences, U.S.A.* **92**, 2–8.
Bousquet, J., Strauss, S. H., Doerksen, A. H. & Price, R. A. (1992a). Extensive variation in evolutionary rate of *rbc*L gene-sequences among seed plants. *Proceedings of the National Academy of Sciences, U.S.A.* **89**, 7844–7848.
Bousquet, J., Strauss, S. H. & Li, P. (1992b). Complete congruence between morphological and *rbc*L-based molecular phylogenies in birches and related species (Betulaceae). *Molecular Biology and Evolution* **9**, 1076–1088.
Bremer, B. & Struwe, L. (1992). Phylogeny of the Rubiaceae and the Loganiaceae-congruence or conflict between morphological and molecular-data. *American*

Journal of Botany **79**, 1171–1184.

Bruneau, A., Dickson, E. E. & Knapp, S. (1995). Congruence of chloroplast DNA restriction site characters with morphological and isozyme data in *Solanum* sect lasiocarpa. *Can. J. Bot.* **73**, 1151–1167.

Chase, M. W., *et al.* (41 others) (1993). Phylogenetics of seed plants – an analysis of nucleotide-sequences from the plastid gene *rbc*L. *Annals of the Missouri Botanical Garden* **80**, 528–580.

Coley, P. D., Bryant, J. P. & Chapin, F. S. I. (1985). Resource availability and plant antiherbivore defense. *Science* **230**, 895–899.

Eggleton, P. & Vane-Wright, R. I. (1994). Some principles of phylogenetics and their implications for comparative biology. *Phylogenetics and ecology.* (ed. P. Eggleton and R. I. Vane-Wright), pp. 345–366. Academic Press.

Feeny, P. P. (1976). Plant apparency and chemical defense. *Recent Advances in Phytochemistry* **10**, 1–40.

Felsenstein, J. (1985). Phylogenies and the comparative method. *American Naturalist* **125**, 1–15.

Foster, S. A. & Janson, C. H. (1985). The relationship between seed size and establishment conditions in tropical woody plants. *Ecology* **66**, 773–780.

Frascaria, N., Maggia, L., Michaud, M. & Bousquet, J. (1993). The *rbc*L gene sequence from chestnut indicates a slow rate of evolution in the Fagaceae. *Genome* **36**, 668–671.

Gittleman, J. L. & Luh, H. K. (1992). On comparing comparative methods. *Annual Review of Ecology and Systematics* **23**, 383–404.

Hancock, J. F. & Pritts, M. P. (1987). Does reproductive effort vary across different life forms and seral environments? A review of the literature. *Bulletin of the Torrey Botanical Club* **114**, 53–59.

Harvey, P. H. (1996). Phylogenies for ecologists. *Journal of Animal Ecology* **65**, 255–263.

Harvey, P. H. & Pagel, M. D. (1991). *The comparative method in evolutionary biology.* Oxford University Press.

Harvey, P. H., Read, A. F. & Nee, S. (1995). Why ecologists need to be phylogenetically challenged. *Journal of Ecology* **83**, 535–536.

Jansen, R. K., Holsinger, K. E., Michaels, H. J. & Palmer, J. D. (1990). Phylogenetic analysis of chloroplast DNA restriction site data at higher taxonomic levels: An example from the Asteraceae. *Evolution* **44**, 2089–2105.

Kelly, C. K. (1995). Seed size in tropical trees: A comparative study of factors affecting seed size in Peruvian angiosperms. *Oecologia* **102**, 377–388.

Kelly, C. K. & Beerling, D. J. (1995). Plant life form, stomatal density and taxonomic relatedness: a reanalysis of Salisbury (1927). *Functional Ecology* **9**, 422–431.

Kelly, C. K. & Purvis, A. (1993). Seed size and establishment conditions in tropical trees – on the use of taxonomic relatedness in determining ecological patterns. *Oecologia* **94**, 356–360.

Kelly, C. K. & Woodward, F. I. (1995). Ecological correlates of carbon isotope composition of leaves: A comparative analysis testing for the effects of temperature, CO_2 and O_2 partial pressures and taxonomic relatedness on delta C-13. *Journal of Ecology* **83**, 509–515.

Kubitzki, K. & Ziburski, A. (1994). Seed Dispersal in Flood Plain Forests of Amazonia. *Biotropica* **26**, 30–43.

Levin, D. A. (1976). Alkaloid-bearing pants: an ecogeographic perspective. *American Naturalist* **110**, 261–284.

Mabberley, D. J. (1987). *The plant book*. Cambridge, Cambridge University Press.

Martins, E. P. & Garland, T. (1991). Phylogenetic analyses of the correlated evolution of continuous characters – a simulation study. *Evolution* **45**, 534–557.

Miles, D. B. & Dunham, A. E. (1993). Historical perspectives in ecology and evolutionary biology – the use of phylogenetic comparative analyses. *Annual Review of Ecology and Systematics* **24**, 587–619.

Mole, S. (1993). The systematic distribution of tannins in the leaves of angiosperms – a tool for ecological studies. *Biochemical Systematics and Ecology* **21**, 833–846.

Patterson, C., Williams, D. M. & Humphries, C. J. (1993). Congruence between molecular and morphological phylogenies. *Annual Review of Ecology and Systematics* **24**, 153–188.

Peat, H. J. & Fitter, A. H. (1994). Comparative analyses of ecological characteristics of British angiosperms. *Biological Reviews of the Cambridge Philosophical Society* **69**, 95–115.

Purvis, A., Gittleman, J. L. & Luh, H. K. (1994). Truth or consequences – effects of phylogenetic accuracy on 2 comparative methods. *Journal of Theoretical Biology* **167**, 293–300.

Purvis, A. & Rambaut, A. (1995). Comparative-analysis by independent contrasts (CAIC): an Apple Macintosh application for analyzing comparative data. *Computer Applications in the Biosciences* **11**, 247–251.

Rees, M. (1995). EC–PC comparative analyses? *Journal of Ecology* **83**, 891–892.

Rice, K. A., Donoghue, M. J. & Olmstead, R. G. (1995). A reanalysis of the large rbcL dataset. *American Journal of Botany* **82** (Supplement), 157–158.

Rodman, J., Price, R. A., Karol, K., Conti, E., Sytsma, K. J. & Palmer, J. D. (1993). Nucleotide-sequences of the rbcL gene indicate monophyly of mustard oil plants. *Annals of the Missouri Botanical Garden* **80**, 686–699.

Salisbury, E. J. (1942). *The reproductive capacity of plants*. London, G. Bell & Sons Ltd.

Westoby, M., Leishman, M. R. & Lord, J. M. (1995). On misinterpreting the 'phylogenetic correction'. *Journal of Ecology* **83**, 531–534.

Willson, M. F. (1983). *Plant reproductive ecology*. New York & Chichester, Wiley & Sons.

Wilson, A. M. & Thompson, K. (1989). A comparative study of reproductive allocation in 40 British grasses. *Functional Ecology* **3**, 297–302.

2 · Phylogenetic uncertainties and sensitivity analyses in comparative biology

Michael J. Donoghue and David D. Ackerly

1 Introduction

In recent years it has become increasingly clear that knowledge of phylogenetic relationships is crucial in extracting historical patterns and possible evolutionary causes from comparative data (e.g. Brooks & McLennan 1991; Harvey & Pagel 1991). Phylogenetic trees provide concrete hypotheses about the chronicle of evolutionary events, including the sequence of splitting events during the evolution of a group and the sequence of character changes (e.g. O'Hara 1988; Donoghue 1989; Maddison & Maddison 1992). Reconstructing character changes helps us avoid trying to explain things that never really happened (Wanntorp 1983), and is necessary in assessing whether changes in different characters were significantly correlated. Although these points are now widely appreciated, there are still few studies of plant ecological traits that have explicitly incorporated phylogenetic trees.

Attempts to put the theory and methods of comparative biology to use raise a wide range of practical issues that need more attention if comparative studies are going to be convincing. In particular, as Harvey & Pagel (1991, pp. 70–71) emphasized, 'Comparative biologists should be aware of the fact that they may well be working with the wrong tree!' It is highly likely, in fact, that virtually every phylogenetic tree found in the literature *is* wrong in one way or another. Does this mean that phylogenetic hypotheses should be ignored? Obviously not! After all, scientists *always* rely on prior inferences that are themselves subject to error. Instead, as Harvey and Pagel (1991, p. 203) rightly concluded, 'Comparative methods need to be developed that take into account the uncertainty about the phylogeny.'

Some attention has been paid recently to these issues, mainly focusing on the use of simulated trees in establishing confidence in particular comparative results (Losos 1994; Martins 1996). Such approaches may be of use when virtually nothing is known at the outset about phylogenetic relationships.

17

Here we focus instead on exploring the implications of a set of proposed and plausible phylogenetic hypotheses, since in practice there are often a relatively small number of alternative topologies deemed worthy of serious consideration. We also emphasize the need for sensitivity tests to address uncertainties that are more directly associated with carrying out a comparative study, especially in scoring taxa for the characters of interest and accounting for taxa that were not actually included in the phylogenetic analysis. Our aim is to provide some practical suggestions for dealing with several sources of phylogenetic uncertainty. In doing so we hope to encourage the development and use of sensitivity analyses to assess the robustness of conclusions derived from comparative studies.

2 Comparative biology

Several of the examples discussed below focus on the evolution of a single character or even a single evolutionary event, and these will seem out of place if 'comparative biology' is equated with particular statistical methods for examining character correlations. This observation compels us to briefly comment on the circumscription of comparative biology, and especially on the distinction made recently between the 'convergence' and 'homology' approaches (for contrasting views, see Coddington 1994; Pagel 1994; Wenzel & Carpenter 1994).

The convergence approach relies on repeated instances of the evolution of a particular kind of characteristic (e.g. Harvey & Pagel 1991). The basic idea is that repeated instances are needed to establish whether there is a significant pattern of association between various traits or between a trait and an environmental variable. This view sees the study of individual evolutionary events, no matter how detailed, as simply lacking the statistical power to establish anything general. The homology approach, in contrast, focuses on the analysis of the circumstances surrounding individual evolutionary changes (e.g., Coddington 1988; Donoghue 1989; Baum & Larson 1991). Under this view, characteristics derived independently in different lineages are not the same (i.e. homologous); or, rather, they are the same only by virtue of having been categorized as such by particular investigators. In any case, it is argued that general patterns observed in multiple lineages will have little bearing on the explanation offered in any particular instance.

This contrast, while perhaps of some heuristic value, will be counter-productive if it tempts us to equate 'real' comparative biology with one approach and dismiss the other. Phylogenetic trees play a central role in both approaches, and each provides valuable and complimentary insights (Coddington 1994). In fact, the most satisfying studies will be those that iterate between approaches. Preliminary comparisons may suggest a set of manipulative experiments, which might in turn suggest a refined adaptive hypothesis, which might then be tested by reference to repeated instances, and so on. The role played by phylogenies is fundamentally the same along the spectrum from 'homology' to increasing 'convergence.' That is, trees allow us to infer whether a particular kind of change occurred once or a number of times, whether change has been in one direction or another, whether change in one character has been associated with change in another, whether associated changes have occurred in a particular sequence, and whether particular changes are associated with shifts in diversification rate (e.g. Sanderson & Donoghue 1994). These are the issues that unite comparative biology, as we conceive it, rather than the use of any particular method (e.g. independent contrasts; Felsenstein 1985).

3 Uncertainties and sensitivity analyses

Ideally one would begin a comparative analysis with perfect knowledge of phylogenetic relationships (including extinct lineages), accurate information on the characters of interest for all populations, and models of character evolution that would provide accurate inferences about character changes on the tree. Obviously, real comparative studies are far from this ideal. We never know the true tree, we have limited knowledge about character distributions, and we lack reliable models of character evolution. For the most part, comparative analyses have ignored such uncertainties, and have instead proceeded by examining a single tree, a single scoring of characteristics, and so on. Whereas this may suffice for demonstration purposes, it leaves nagging doubts about the robustness of the results. How, one wonders, would the results differ with a somewhat different tree, or scoring, or evolutionary model?

Here we will mainly consider uncertainties and sensitivity analyses relating to tree topology, drawing on examples from our own work, and with an emphasis on broad analyses of angiosperms. We also touch briefly on several uncertainties that arise in coding characters and in trying to accommodate mismatches between the available comparative data and the available phylogenetic information. Simulation studies of the independent contrasts method have examined the effect of choosing different models of character evolution (Martins & Garland 1991; Diaz-Uriarte & Garland 1996), and we will not address this problem explicitly here. Ours is by no means an exhaustive treatment of the possible sources of uncertainty or sensitivity tests; instead, we have tried to provide ideas on how to proceed, which might then be adapted to the circumstances surrounding any particular comparative study.

3.1 Uncertainties about tree topology

Estimates of phylogeny may be erroneous for a variety of reasons, only a few of which are mentioned here. First, and most obviously, the available data may be too few or too noisy to yield an accurate estimate of relationships. Second, a perfectly accurate gene tree might not reflect phylogenetic relationships among species, owing to hybridization, lineage sorting and/or lateral transfer (Pamilo & Nei 1988; Doyle 1992; Clark *et al.* 1994; Maddison 1995). Third, estimation methods (e.g. parsimony, maximum likelihood, etc.) may be statistically inconsistent under some evolutionary circumstances (e.g. high and uneven rates of change), such that, in the worst cases, the addition of data leads to greater confidence in the wrong relationships (e.g. Felsenstein 1978; Penny *et al.* 1992; Huelsenbeck & Hillis 1993). Fourth, even when methods are consistent, optimal solutions may not have been found. This problem relates to the fact that many phylogenetic problems are computationally challenging (Maddison *et al.* 1992; Rice *et al.* 1995, in prep.; Swofford *et al.* 1996). For example, exact solutions cannot generally be obtained for parsimony problems involving more than 20 or 30 taxa. Instead, larger problems rely on heuristic search algorithms (generally involving some form of branch swapping), and some or all optimal solutions may not be obtained owing to the limitations of hill-climbing algorithms (finding local, not global, optima) and practical constraints on computer time.

A pertinent example of computational limitations (and perhaps all four problems) is provided by the preliminary analysis of 500 seed plant *rbcL* sequences conducted by Chase *et al.* (1993). In the case of their B-series trees (judged by them to be the more reliable), searches were conducted for approximately one month on a Macintosh Quadra computer using PAUP (Swofford 1993), resulting in the discovery of 3900 trees of 16 538 steps (all characters included). A reanalysis of this dataset by Rice *et al.* (1995, in prep.; http://green.harvard.edu/ ~ rice/treezilla), running on several SUN workstations for a total of approximately nine months, yielded many shorter trees, including 8975 trees five steps shorter than the published trees. Longer searches would undoubtedly yield even shorter trees.

Comparative studies based on the Chase *et al.* (1993) analysis have tended to use the single published B-series tree (e.g., Silvertown & Dodd, this volume). The consequences of this choice have not been examined. As a first step in this direction we have conducted a simple simulation study to explore the differences between the single Chase *et al.* tree, the 7670 fully resolved most parsimonious trees found by Rice *et al.* (1995; 1305 trees have polychotomies due to collapsed zero-length branches), and a set of 100 trees of 500 taxa generated using MacClade's random tree function (Maddison & Maddison 1992; see Martins 1996, for a thorough discussion of random tree generation).

Our simulations consisted of the following steps. First, one fully resolved tree was arbitrarily chosen from among the most parsimonious trees of Rice *et al.* to be the 'true tree' (we used the single tree available on the treezilla web site). Second, we simulated the evolution of two continuous characters on this topology, starting from the root and proceeding upward to the 500 terminal taxa. At each node, the changes in the two characters along each daughter branch were selected at random from a bivariate normal distribution with correlation coefficients (C_I; = the 'input correlation' of Martins & Garland 1991) of 0 and 0.5 in our two simulation runs. This corresponds to a 'speciational' model of character evolution in which expected change is independent of branch length.

From the resulting character states we then calculated: (a) the ahistorical correlation (C_A; 'tip correlation' of Martins & Garland 1991) based on the character states across all 500 taxa; (b) the 'observed' historical correlation (C_O) based on independent contrasts calculated from actual character values during the simulation; and (c) the reconstructed historical correlation (C_R), calculated by inferring the history of character change using the squared change parsimony algorithm of Maddison (1991), and again calculating the correlation coefficient from the independent contrasts at each node. The use of squared change parsimony differs from methods employed in other implementations of the independent contrasts approach (e.g., Martins & Garland 1991; Purvis & Rambaut 1995), and the effect of this difference has not been systematically investigated (cf. Diaz-Uriarte & Garland 1996). The program used to carry out these simulations (available from D. Ackerly) accepts trees coded in standard NEXUS-format parenthetical notation, which may be especially useful in carrying out sensitivity analyses over many trees.

As expected, the observed and reconstructed evolutionary correlations on the 'true tree' were virtually identical in each of the two simulations, and these two parameters were also fairly close to the input correlations (simulation A:

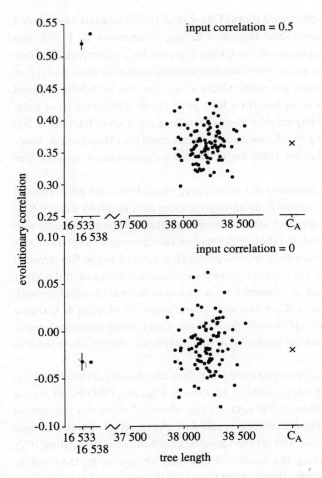

Figure 2.1. Analyses of simulated character evolution on alternative phylogenetic trees from 500 taxon *rbc*L analyses of seed plants. The reconstructed historical correlation for the two characters is plotted against tree length for 7670 fully resolved *rbc*L trees of 16 533 steps, the published tree from Chase *et al.* (1993) of 16 538 steps, and 500 random trees generated by MacClade (Maddison & Maddison 1992). On the right is the ahistorical correlation (C_A), calculated using data from the terminal taxa. See text for details.

$C_I = 0.5$, $C_O = 0.536$, $C_R = 0.519$, $C_A = 0.363$; B: $C_I = 0.0$, $C_O = -0.021$, $C_R = -0.032$, $C_A = -0.020$). Comparison of the ahistorical and historical correlations (especially for simulation A), illustrate how different these can be, even when large numbers of taxa are considered, illustrating the utility of the independent contrasts method. Of most importance for the present discussion, the results over the 7670 most parsimonious *rbc*L trees were clustered very tightly around the value observed for the one 'true tree', with a total range of less than 0.02 in both simulations, and the correlations observed for the slightly less parsimonious tree of Chase *et al.* (1993) were similar to these in both cases (Figure 2.1). We do not know whether the relative consistency in the outcomes that we observed here will hold for other characters or other types of analyses; in fact, we expect that it will not hold in some cases. *The sensitivity of comparative analyses to topological variation must therefore be examined on a case-by-case basis, as we have done here, by evaluating the set of relevant trees.*

The historical correlations calculated for the set of 100 random trees differed in two important respects from the results for the *rbc*L trees (Figure 2.1). First, we found extensive variation among the random trees. This demonstrates that there are alternative phylogenies that will lead to markedly different conclusions. Second, the distribution of correlations calculated on the random trees centers around the ahistorical correlation (also see the *Acer* example below). Thus, the mean of the historical correlations calculated from a set of random trees apparently does not provide a provisional estimate of the historical correlation, as suggested by Martins (1996). At least a preliminary phylogenetic hypothesis is needed for this purpose. Analyses using random trees do, however, set bounds on the possible results of a comparative analysis (Losos 1994).

3.2 Uncertainties about rooting

Uncertainty concerning the exact placement of the root of a tree is common in phylogenetic studies. In part this may be due to the relatively large number of evolutionary changes separating any ingroup taxa and possible outgroups, and the consequent effects of homoplasy, which may render a number of different rootings equally or almost equally parsimonious (e.g. Felsenstein 1978; Maddison *et al.* 1984; Wheeler 1990; Donoghue 1994).

One well known case of uncertainty concerns the position of the root of the angiosperm tree (see Doyle & Donoghue 1993; Crane *et al.* 1995; Taylor & Hickey 1996). Although different datasets are in considerable agreement about a number of major clades within angiosperms (e.g. monocots, eudicots, etc.), different analyses support a woody 'magnoliid' rooting (e.g. Soltis *et al.* 1996), a 'paleoherb' rooting (e.g. Doyle *et al.* 1994), or in the case of the *rbc*L analyses, a *Ceratophyllum* rooting (e.g. Chase *et al.* 1993). Under these circumstances, how should comparative studies proceed? One possibility is to simply explore the consequences of the different plausible topologies for the question of interest. In connection with the simulations described above, we also calculated correlations on four angiosperm rootings designed to mimic viable alternatives to the *Ceratophyllum* rooting found in *rbc*L trees, including woody magnoliid and paleoherb options. These alternative topologies yielded correlations virtually identical to those found for the 'true tree', indicating that root placement has little impact on the method of independent contrasts in this particular case.

This kind of sensitivity analysis has been carried out in other studies involving the angiosperm tree. Sanderson & Donoghue (1994) and Weller *et al.* (1995) examined the effect of alternative rootings on, respectively, rates of diversification and self-compatibility, and found no significant differences among these. In contrast, interpretation of the evolution of many other angiosperm characters depends directly on which rooting is chosen (Doyle & Donoghue 1993). For example, whether the first angiosperms are inferred to have been woody or herbaceous plants depends on the choice between a magnoliid or paleoherb rooting. Similarly, interpretation of a variety of flower characters (e.g. many versus few flower parts, spiral versus whorled arrangement of parts) depends ultimately on resolution of the rooting problem. In general, *the impact of alternative rootings will need to be established on a case-by-case basis.*

3.3 Neighbouring trees

The above discussion might suggest that sensitivity analyses are advised only when there are alternative equally parsimonious trees from a single analysis, or alternative trees from different analyses of the same problem. However, testing the robustness of results seems wise even when only one optimal phylogenetic hypothesis has been identified. In particular, we suggest examining what happens to a correlation as one backs away from an optimal tree, in order to assess just how strongly the conclusions of a comparative analysis hinge on commitment to the most parsimonious tree. This approach is illustrated by a preliminary study of the evolution of branching architecture in species of *Acer* (D. Ackerly & M. Donoghue, unpublished data).

An analysis of combined morphological and molecular data for seven species of *Acer* yielded a single most parsimonious tree of length 233, as well as a single tree of length 234, three trees of length 235, four of 236, and so on. Of the 10 395 possible rooted bifurcating trees for seven taxa, 398 fall within

20 steps of the most parsimonious tree, and these trees were saved for use in comparative analyses. In addition, 500 random trees were generated (using MacClade; Maddison & Maddison 1992) to examine the range of possible results. Using these trees, we tested the historical correlation between the rate of terminal and lateral branch growth, traits that influence sapling regeneration in relation to light environment (cf. Sakai 1987). These were measured in the field on saplings and on the branches of adult trees for all seven species in the phylogenetic analysis.

Terminal and lateral branch growth have a strong negative historical correlation, but a much weaker ahistorical correlation ($C_R = -0.747$; $C_A = -0.352$; Figure 2.2). The negative correlation inferred from the most parsimonious tree is generally upheld on trees up to ten steps longer, at which point correlations near zero were observed for some trees. The results for 500 random trees (shown as open symbols in Figure 2.2) illustrate the magnitude of variation among other conceivable trees. As in the *rbc*L example above, the mean of the results for the random trees is almost identical to the ahistorical correlation.

Inspection of the trees revealed that the abrupt changes that occur in trees longer than 267 steps is due to the loss of two clades, which contributed strongly to the negative correlation among contrasts (Figure 2.3). Changes of this sort are not limited to analyses of small numbers of taxa, as can be seen in the *rbc*L case above. This observation highlights the desirability of investigating the strength of support for clades (as judged, for example, by bootstrap or decay analyses) that may be especially significant from the standpoint of the comparative analysis. In studies involving many taxa it may not be feasible to examine all trees in the neighborhood of the optimal trees, but it may be possible to design sensitivity tests that focus on clades that are identified in the original phylogenetic analysis as being especially weakly supported.

3.4 Uncertainties in scoring

A variety of uncertainties tend to arise in designing comparative analyses owing to mismatches between the data available on the characters of interest and the available phylogenetic trees. So far, such practical issues have attracted very little attention. Here we merely highlight several such problems and possible sensitivity tests.

Perhaps the most common difficulty is the lack of relevant character information for taxa included in the available phylogenies (e.g. missing information on mode of dispersal in fossil taxa; Donoghue 1989). These taxa can

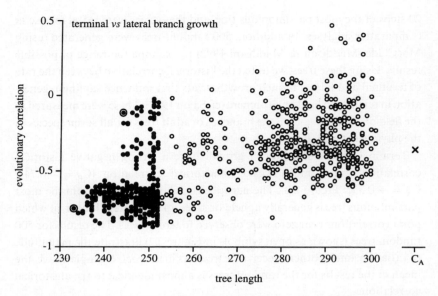

Figure 2.2. Reconstructed historical correlation between rates of terminal and lateral branch growth, plotted against tree length for all trees within 20 steps of the most parsimonious tree (filled symbols), and for 500 random trees generated in MacClade (open symbols). See text for details. The two circled points represent the trees illustrated in Figure 2.3.

be coded as 'missing' in character optimizations, contrast analyses, etc., but one wonders how greatly the results of an analysis might change if character information became available. A related problem concerns variation or polymorphism within terminal taxa, which may be more common in characters investigated by evolutionary ecologists (as opposed to systematists interested in higher level relationships), especially as extensive population-level information may have been gathered. In addition, a single species or other terminal taxon in a tree may represent a larger clade, and it may be tempting to score it as polymorphic as a means of expressing the understanding that relatives differ in state from the taxon that happened to be included in the phylogenetic analysis. Again, one wonders what would happen if polymorphisms were resolved in one way or another.

It is important to recognize that the problems posed by missing and polymorphic characters are somewhat different in a comparative analysis than in a phylogenetic analysis (see Nixon & Davis 1991; Platnick et al. 1991; Maddison & Maddison 1992; Donoghue 1994), and some solutions are more appealing in the context of a comparative study. In particular, it may be

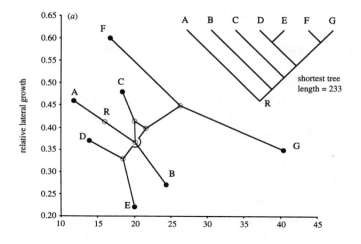

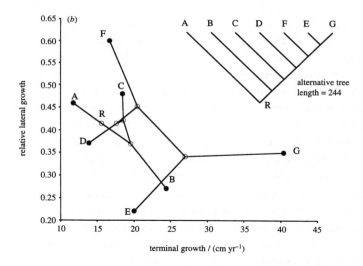

Figure 2.3. Scatterplot of terminal versus lateral branch growth for seven species of maple (filled symbols), and reconstructed values for ancestral taxa based on squared change parsimony (open symbols). The phylogeny, as shown on the right, is mapped onto the character space. (a) The most parsimonious tree, which results in an evolutionary correlation between the two traits of − 0.75. (b) A tree 11 steps longer with a different pattern of character change and smaller evolutionary correlation of − 0.11. The change in the correlation is due to changed relationships among species D, E, F and G; in (a) clades DE and FG contributed large negative assocations to the overall correlation, while in (b) clades DF and EG contribute positive associations.

desirable to split up terminal taxa into two or more attached branches on the basis of the character data at hand (cf. Pagel 1992; Purvis & Rambaut 1995). It must then be appreciated, however, that comparative tests are biased conservatively by virtue of having minimized evolutionary change within terminal taxa (i.e. there may have been more changes in the true phylogeny).

A variety of these character coding issues were encountered in the Weller *et al.* (1995) study of self-incompatibility. Alternative resolutions of missing and polymorphic data were explored on a wide range of angiosperm trees. In most cases self-compatibility was optimized as ancestral for angiosperms, and even in those few cases where self-incompatibility was ancestral it was always most parsimonious to suppose that not all self-incompatibility systems were homologous (retained from the ancestral condition); instead several independent losses and originations were required. A similar approach has been applied in a study of the evolution of dioecy in monocotyledons (G. Weiblen & M. Donoghue, in prep.). In this case several codings of a three-state breeding system character (hermaphroditic flowers, dioecy, monoecy) were investigated, including consistently scoring all polymorphisms in favour of each one of the three states. Based on each scoring, the number of transitions between states was then determined on a composite phylogeny of monocots as a means of placing bounds on the number of changes in each direction. Despite considerable polymorphism, the range in the number of inferred changes was not great and clear patterns of transition bias emerged. For example, changes from dioecy to any other state were found to be exceptionally rare regardless of how polymorphisms were resolved.

It should be noted that character coding experiments such as those described above do not explore all possible permutations and combinations of possible scorings, and as such may not provide an accurate indication of the bounds on the possible outcome. Nevertheless, such exercises may provide a useful expedient when the number of possible resolutions of missing and polymorphic scores is very large. Another possibility would be to repeatedly assign single states at random to those taxa scored as missing or polymorphic, and generate a range of correlations from these alternative scorings (similar approaches apply for continuous variables).

Another critical element of character analysis concerns the weighting of character state changes. In practice, transitions between states (or shifts in different directions in the case of continuous variables) are generally treated as though they are equally likely to occur. That is, a transition from state A to B entails the same cost as from B to A. It would be useful to explore the sensitivity of comparative results to changes in this standard assumption. This can be done using a series of step matrices (e.g. in MacClade; Maddison

& Maddison 1992; Maddison 1994) designed to reflect possible inequalities in transitions among states. In each case, the effect of differential weighting on a particular optimization or correlation could be recorded so as to determine the range of weights over which a particular outcome holds. In our laboratory (MJD), this approach has been used in pilot studies of the evolution of fruit types and zygomorphic flowers in Asteridae (unpublished data). In both cases, confidence in the outcome based on equal weighting of state changes is bolstered by the observation that many of the same results are obtained over a reasonably large range of alternative weighting schemes.

3.5 Uncertainties in adding taxa

Another common source of uncertainty in comparative analyses concerns the addition of taxa that were not included in the underlying phylogenetic analyses. In some cases this may involve one or a few species for which comparative data are available, and in other cases it may be tempting to add entire clades. Clearly, such procedures entail significant risks, and the robustness of the conclusions to changes in assumed relationships could be tested in a variety of ways. Most obviously, comparative results can be conducted with and without the additional taxa to determine whether a particular result hinges on the added information. Second, the position of an added species can be shifted to a variety of alternative positions in the tree, to see whether this makes a difference in the outcome. This procedure may require considerable understanding of previous taxonomic treatments, all of which may be misguided about relationships.

The addition of whole trees obtained from separate analyses is appealing from the standpoint of increased sample size, and has become a common practice (e.g. Donoghue 1989; Sillen-Tullberg 1993; Hoglund & Sillen-Tullberg 1994). However, it must be appreciated that this may result in a composite phylogeny that is not a globally parsimonious solution. That is, a simultaneous analysis of all of the taxa could yield a significantly different topology. In effect, the efficacy of the process of piecing together separately derived trees rests on the strength of support for assumptions about the monophyly of particular groups. For example, in an analysis of the historical relationship between dioecy and fleshy propagules, Donoghue (1989) assembled a composite phylogeny by adding an angiosperm tree and a conifer tree to an underlying analysis of relationships among major lines of seed plants. This assumes that both angiosperms and conifers are monophyletic groups, that the position of these clades and other major lines of seed plants would not shift in a simultaneous analysis, and that relationships within each group would not be altered in a global phylogenetic study. The strength of such assumptions is clearly variable. Thus, whereas angiosperm monophyly has been amply confirmed (Doyle & Donoghue 1993; Crane *et al.* 1995), conifer monophyly is less certain (Chase *et al.* 1993; Rothwell & Serbet 1994).

The obvious solution to these problems is to carry out a global phylogenetic analysis, but this will often not be possible. On the one hand, the data from different studies may not be readily combinable into one matrix, and on the other hand, this might create the computational difficulties associated with large datasets. In this case we can only suggest that the sensitivity of comparative analyses might be checked by consideration of a set of plausible alternative composite phylogenies. Steps may also need to be taken to avoid biases in comparative analyses that may result from joining together a possibly biased selection of phylogenies (O'Hara 1992; Sillen-Tullberg 1993; Hoglund & Sillen-Tullberg 1994).

4 TreeBASE

We have emphasized the desirability of taking into account alternative phylogenetic hypotheses. However, we certainly appreciate that this tends to be easier said than done, in part because phylogenetic information has not been readily accessible. The need to develop tools to improve access to phylogenetic knowledge has been recognized (e.g. Sanderson *et al.* 1993; Blake *et al.* 1994; Donoghue 1994), and a prototype relational database of phylogenetic information – TreeBASE – has been developed with this end in mind (Sanderson *et al.* 1994; http://phylogeny.harvard.edu/treebase).

The prototype versions of TreeBASE now contain 155 phylogenetic studies of green plants (about 3500 taxa, 410 trees), and data matrices are present for about 90% of these studies. TreeBASE allows for browsing and searching on authors, key words, and taxonomic names and provides tools for downloading datasets, as well as a submission form for receipt of additional studies. The goal is to include phylogenetic data on all groups of organisms, and we anticipate that journals will eventually require electronic submission of trees and data matrices as a corequisite of publication.

5 Conclusions

Phylogenetic studies of plant life history characteristics are certainly promising, but if such studies are going to be convincing it will be necessary to deal with a variety of phylogenetic uncertainties. Our focus has been on cases in which a set of plausible phylogenetic hypotheses exists at the outset of a study, as is often the case in angiosperms. Tree-based methods have been developed for cases in which nothing is known about relationships for all or part of a tree (Losos 1994; Martins 1996). Judging by polychotomies in published trees, this may seem commonplace, but these do not necessarily signify that all possible resolutions are equally well supported (see Maddison 1989), and it may be best to consider each of the equally supported alternative topologies rather than resort to random trees. Even using the characters under study to help constrain a phylogenetic hypothesis might be justified (cf. Pagel 1992; Purvis & Rambaut 1995), either from the standpoint of consciously providing a conservative test (de Queiroz 1996) or providing a better estimate of the historical correlation than the mean of a set of random trees (see above).

At this early stage in the development of our understanding of plant relationships (Donoghue 1994), it is clearly not wise to become narrowly focused on a few phylogenetic hypotheses. While it is evidently tempting to rely on a single phylogenetic analysis, such as the study of *rbc*L sequences presented by Chase *et al.* (1993), we have already achieved a richer understanding of the phylogeny of most groups than is portrayed in such broad analyses, and the more detailed hypotheses for particular groups may be most appropriate for many comparative studies. We appreciate the theoretical difficulties in piecing together more detailed phylogenetic trees, but the benefits may be great and the risks involved might be less than reliance on a single (perhaps suboptimal) tree. Broad phylogenetic analyses also tend to include a limited and biased sample of taxa, and the intercalation of additional taxa for purposes of a comparative analysis entails its own set of risks.

It would also be unfortunate to focus on just a few comparative methods. In particular, despite the popularity and power of phylogenetically independent contrasts (Felsenstein 1985; Harvey & Pagel 1991), there is much to be learned from the analysis of discrete characters (Maddison 1990), or mixtures of discrete and continuous variables (cf. Purvis & Rambaut 1995). Furthermore, independent contrast methods, as these are usually implemented, are unable to keep track of the order of evolutionary events (but see McPeek 1995). Inasmuch as establishing such sequences is often critical in testing causal theories about character evolution (e.g. O'Hara 1988; Donoghue 1989), this may be a very significant limitation. Finally, as emphasized above, deep understanding of the causes of evolutionary change and ecological patterns is most likely to emerge from the integration of 'homology' and 'convergence' approaches, and it seems counterproductive to restrict the purview of 'comparative biology' to one or the other.

We thank J. Silvertown and the other organizers of the Royal Society symposium for inviting us to participate, and D. Baum, W. Maddison, W. Piel, K. Rice, M. Sanderson, G. Weiblen, and members of Harvard's Bio 216 seminar for helpful discussion of these issues. Development of TreeBASE was supported by a US National Science Foundation grant to MJD (DEB-9318325). DDA was supported by the Arnold Arboretum and a NSF postdoctoral fellowship (DEB-9403252).

References

Baum, D. A. & Larson, A. (1991). Adaptation reviewed: a phylogenetic methodology for studying character macroevolution. *Systematic Zoology* II **40**, 1–18.

Blake, J. A., Bult, C. J., Donoghue, M. J., Humphries, J., & Fields, C. (1994). Interoperability of biological databases: a meeting report. *Systematic Biology* **42**, 562–568.

Brooks, D. R. & McLennan, D. A. (1991). *Phylogeny, ecology, and behavior: a research program in comparative biology.* University of Chicago Press.

Chase, M. W., *et al.* (41 others). (1993), Phylogenetics of seed plants: an analysis of nucleotide sequences from the plastid gene *rbc*L. *Annals of the Missouri Botanical Garden* **80**, 528–580.

Clark, J. B., Maddison, W. P. & Kidwell, M. G. (1994). Phylogenetic analysis supports horizontal transfer of *P* transposable elements. *Molecular Biology and Evolution* **11**, 40–50.

Coddington, J. A. (1988). Cladistic tests of adaptational hypotheses. *Cladistics* **4**, 3–22.

Coddington, J. A. (1994). The roles of homology and convergence in studies of adaptation. In *Phylogenetics and ecology* (ed. P. Eggleton & R. Vane-Wright), pp. 53–78. London: Academic Press.

Crane, P. R., Friis, E. M. & Pederson, K. R. (1995). The origin and early diversification of angiosperms. *Nature* **374**, 27–33.

de Queiroz, K. (1996). Including the characters of interest during tree reconstruction and the problems of circularity and bias in studies of character evolution. *American Naturalist* **148**, 700–708.

Diaz-Uriarte, R. & Garland Jr., T. (1996). Testing hypotheses of correlated evolution using phylogenetically independent contrasts: sensitivity to deviations from Brownian motion. *Systematic Biology* **45**, 27–47.

Donoghue, M. J. (1989). Phylogenies and the analysis of evolutionary sequences, with examples from seed plants. *Evolution* **43**, 1137–1156.

Donoghue, M. J. (1994). Progress and prospects in reconstructing plant phylogeny. *Annals of the Missouri Botanical Garden* **81**, 405–418.

Doyle, J. A. & Donoghue, M. J. (1993). Phylogenies and angiosperm diversification. *Paleobiology* **19**, 141–167.

Doyle, J. A., Donoghue, M. J. & Zimmer, E. A. (1994). Integration of morphological and ribosomal RNA data on the origin of angiosperm. *Annals of the Missouri Botanical Garden* **81**, 419–450.

Doyle, J. J. (1992). Gene trees and species trees: molecular systematics as one-character taxonomy. *Systematic Botany* **17**, 144–163.

Felsenstein, J. (1978). Cases in which parsimony or compatibility methods will be positively misleading. *Systematic Zoology* **27**, 401–410.

Felsenstein, J. (1985). Phylogenies and the comparative method. *American Naturalist* **125**, 1–15.

Harvey, P. H. & Pagel, M. D. (1991). *The comparative method in evolutionary biology*. Oxford University Press.

Hoglund, J. & Sillen-Tullberg, B. (1994). Does lekking promote the evolution of male-biased size dimorphism in birds? On the use of comparative approaches. *American Naturalist* **144**, 881–889.

Huelsenbeck, J. P. & Hillis, D. M. (1993). Success of phylogenetic methods in the four-taxon case. *Systematic Biology* **42**, 247–264.

Losos, J. B. (1994). An approach to the analysis of comparative data when a phylogeny is unavailable or incomplete. *Systematic Biology* **43**, 117–123.

Maddison, D. R. (1994). Phylogenetic methods for inferring the evolutionary history and processes of change in discretely valued character. *Annual Review of Entomology* **39**, 267–292.

Maddison, D. R., Ruvolo, M. & Swofford, D. L. (1992). Geographic origins of human mitochondrial DNA: phylogenetic evidence from control region sequences. *Systematic Biology* **41**, 111–124.

Maddison, W. P. (1989). Reconstructing character evolution on polytomous cladograms. *Cladistics* **5**, 365–377.

Maddison, W. P. (1990). A method for testing the correlated evolution of two binary characters: are gains or losses concentrated on certain branches of a phylogenetic tree? *Evolution* **44**, 539–557.

Maddison, W. P. (1991). Squared-change parsimony reconstructions of ancestral states for continuous-valued characters on a phylogenetic tree. *Systematic Zoology* **40**, 304–314.

Maddison, W. P. (1995). Phylogenetic histories within and among species. In *Experimental and molecular approaches to plant biosystematics* (ed. P. Hoch & A. Stephenson), pp. 273–287. St. Louis: Missouri Botanical Garden.

Maddison, W. P., Donoghue, M. J. & Maddison, D. R. (1984). Outgroup analysis and parsimony. *Systematic Zoology* **33**, 83–103.

Maddison, W. P. & Maddison, D. R. (1992). *MacClade: interactive analysis of phylogeny and character evolution*. Sunderland, Massachusetts: Sinauer.

Martins, E. P. (1996). Conducting phylogenetic comparative studies when the phylogeny is not known. *Evolution* **50**, 12–22.

Martins, E. P. & Garland Jr., T. (1991). Phylogenetic analyses of the correlated evolution of continuous characters: a simulation study. *Evolution* **45**: 534–557.

McPeek, M. A. (1995). Testing hypotheses about evolutionary change on single branches of a phylogeny using evolutionary contrasts. *American Naturalist* **145**, 686–703.

Nixon, K. C. & Davis, J. I. (1991). Polymorphic taxa, missing values and cladistic analysis. *Cladistics* **7**, 233–241.

O'Hara, R. J. (1988). Homage to Clio, or, toward an historical philosophy for evolutionary biology. *Systematic Zoology* **37**, 142–155.

O'Hara, R. J. (1992). Telling the tree: narrative representation and the study of evolutionary history. *Biol. Phil.* **7**, 135–160.

Pagel, M. D. (1992). A method for the analysis of comparative data. *Journal of Theoretical Biology* **156**, 431–442.

Pagel, M.D. (1994). The adaptationist wager. In *Phylogenetics and ecology* (ed. P. Eggleton & R. Vane-Wright), pp. 29–51. London: Academic Press.

Pamilo, P. & Nei, M. (1988). Relationships between gene trees and species trees. *Molecular Biology and Evolution* **5**, 568–583.

Penny, D., Hendy, M. D. & Steele, M. A. (1992). Progress with methods for constructing evolutionary trees. *Trends in Ecology and Evolution* **7**, 73–79.

Platnick, N. I., Griswold, C. E. & Coddington, J. A. (1991). On missing entries in cladistic analysis. *Cladistics* **7**, 337–343.

Purvis, A. & Rambaut, A. (1995). Comparative analysis by independent contrasts (CAIC): an Apple Macintosh application for analysing comparative data. *Computer Applications in the Biosciences* **11**, 247–251.

Rice, K. A., Donoghue, M. J. & Olmstead, R. G. (1995). A reanalysis of the large *rbc*L dataset. *American Journal of Botany* **82**, S157–158 (abstract).

Rothwell, G. R. & Serbet, R. (1994). Lignophyte phylogeny and the evolution of spermatophytes: a numerical cladistic analysis. *Systematic Botany* **19**, 443–482.

Sakai, S. (1987). Patterns of branching and extension growth of vigorous saplings of Japanese *Acer* species in relation to their regeneration strategies. *Canadian Journal of Botany* **65**, 1578–1585.

Sanderson, M. J., Baldwin, B. G., Bharathan, G., et al. (1993). The growth of phylogenetic information and the need for a phylogenetic database. *Systematic Biology* **42**, 562–568.

Sanderson, M. J. & Donoghue, M. J. (1994). Shifts in diversification rate with the origin of angiosperms. *Science* **264**, 1590–1593.

Sanderson, M. J., Donoghue, M. J., Piel, W. & Eriksson, T. (1994). TreeBASE: a prototype database of phylogenetic analyses and an interactive tool for browsing the phylogeny of life. *American Journal of Botany* **81**, S183 (abstract).

Sillen-Tullberg, B. (1993). The effect of biased inclusion of taxa on the corelation of discrete characters in phylogenetic trees. *Evolution* **47**, 1182–1191.

Soltis, D. E., Soltis, P. S., Nickrent, D. L., Johnson, L. A., et al., (1996). Phylogenetic relationships among angiosperms inferred from 18S rDNA sequences. *Proceedings of the National Academy of Sciences, USA.* (In the press).

Swofford, D. L. (1993). *PAUP: phylogenetic analysis using parsimony, version* 3.1.1. Washington, DC: Smithsonian Institution.

Swofford, D. L., Olsen, G. J., Waddell, P. J. & Hillis, D. M. (1996). Phylogenetic inference. In *Molecular systematics*, second edition (ed. D. Hillis, C. Moritz, & B. Mable), pp. 407–514. Sunderland, Massachusetts: Sinauer.

Taylor, D. W. & Hickey, L. J. (ed.). (1996). *Flowering plant origin, evolution and phylogeny.* New York: Chapman & Hall.

Wanntorp, H.-E. (1983). Historical constraints in adaptation theory: traits and non-traits. *Oikos* **41**, 157–159.

Weller, S. G., Donoghue, M. J. & Charlesworth, D. (1995). The evolution of self-incompatibility in angiosperms: a phylogenetic approach. In *Experimental and molecular approaches to plant biosystematics* (eds P. Hoch & A. Stephenson), pp. 355–382. St. Louis: Missouri Botanical Garden.

Wenzel, J. W. & Carpenter, J. M. (1994). Comparing methods: adaptive traits and tests of adaptation. In *Phylogenetics and ecology* (eds P. Eggleton & R. Vane-Wright), pp. 79–101. London: Academic Press.

Wheeler, W. C. (1990). Nucleic acid sequence phylogeny and random outgroups. *Cladistics* **6**, 363–367.

3 · Comparative ecology of the native and alien floras of the British Isles

M.J. Crawley, P.H. Harvey and A. Purvis

1 Introduction

What are the ecological attributes associated with successful invasion of new habitats? At first glance, it might seem that any search for traits associated with invasive ability would be bound to fail. All species in their native habitats must exhibit the ability to increase when rare (the 'invasion criterion', $dN/dt > 0$ when N is small). If this were not the case, then a species would drift inexorably downwards towards extinction, as one environmental calamity followed another. Since all species must pass the invasion criterion, it is evident that *all* species possess the traits necessary for invasion. It is equally clear that 'competitive ability' is not a species-specific trait; it depends upon the identities of the species with which a plant is competing and upon the environmental conditions under which competition takes place. Likewise a species which was less competitive in one year might be more competitive in a different kind of year (Crawley 1989, 1990). Given these constraints, we would predict that different traits would be associated with different successional stages (e.g. r-strategists with rapid development, wide seed dispersal in space or time and microsite-limited recruitment, *vs* K-strategists with large size, long life and seed-limited recruitment), different habitats (e.g. forests *vs* grasslands), and different resource supply rates (e.g. nitrogen-rich *vs* nitrogen-poor soils). We might expect, therefore, that a minimal statistical description of the traits of successful alien plants would involve a high-dimensional model, requiring specification of the abiotic environment, the abundance and specific identity of competitors, herbivores, mutualists and natural enemies, the successional stage, the disturbance regime and the idiosyncrasies of the year in which the attempted invasion occurs. In this chapter, we step back from this detailed perspective to take the broadest possible view. We ask which traits distinguish the native and alien members of the British flora. The question of what distinguishes successful from unsuccessful invaders is left for another occasion.

2 Methods

2.1 Definitions

Natives: plant species that would be present without human intervention (in the case of the British Isles, this means those plants which returned during the first 10 000 years after the retreat of the glaciers before humans began mass transport of plants).

Aliens: species introduced by humans after 500 BC. The date is arbitrary, but it allows that recorded Roman introductions are classed as aliens. It also means that many plants associated with human habitats and agricultural fields are classed as natives (much as the house sparrow *Passer domesticus* is classed as a native bird). Continental botanists refer to these ancient, human-associated plants as archaeophytes.

Naturalized: aliens with self-replacing populations. For the purposes of this study, the naturalized aliens are defined as those species treated as such by Kent (1992) in his *List of Vascular Plants of the British Isles*. The classification of many species as native or alien, or of aliens as naturalized or not, is highly contentious. This can be seen by comparing the opinions of different modern authors like Stace (1991) in his *New Flora of the British Isles* and Clement & Foster (1994) in their *Alien Plants of the British Isles*.

Casuals: alien plant species which do not form self-replacing populations, and rely on repeated reintroduction for any semblance of permanence that they might exhibit.

Introduction: the transport (intentional or unintentional) of seeds or viable plant parts from one country to a habitat in another country where the species did not previously occur.

Establishment: the formation of a self-replacing population. Many long-lived introductions (e.g. ornamental trees) give the appearance of establishment but never produce a second generation; such species are best viewed as long-lived casuals.

2.2 Data sources

The identities of the native and alien species are taken from Kent (1992). Information on their phylogeny is extracted from Chase *et al.* (1993) supplemented where necessary by the taxonomy used in Stace (1991). Country of origin and mode of introduction were obtained from a wide range of sources, as detailed by Crawley (1997). Distribution and abundance of aliens within British habitats were obtained from more than 100 county floras and from standard works (e.g. Stace 1991; Clapham *et al.* 1962). Data on plant traits come mainly from Grime *et al.* (1988) supplemented with information from the ecological flora database (Fitter & Peat 1994). The analysis of plant traits excluded ferns and horsetails (too little data), but included gymnosperms and a few angiosperm hybrids (where these had a native and an alien parent) and subspecies (where distinctive native and alien members of the same species were identified by Kent, (1992)).

2.3 Analysis

The need to consider phylogenetic relatedness in comparative studies is widely accepted. At first sight, it might appear that our study provides an exception to this rule. Given that more or less every invader to the U.K. invaded independently, can species be treated as statistically independent? The answer is that they cannot. Aliens are not a phylogenetically random subset of the British flora (e.g. Pinaceae has a high proportion of aliens and Cyperaceae has a low proportion; see Crawley 1997). Because other attributes of organisms obviously tend to covary with phylogeny, aliens will not constitute a random subset of the British flora with respect to these traits either. We should not therefore be surprised to find that attributes, and correlations among them, differ significantly between aliens and natives when species are taken as independent points (for further discussion see Harvey & Pagel 1991; Harvey *et al.* 1995; Rees 1995).

We have used two comparative methods, both of which map the data from different species on to a phylogeny in order to partition the among-species variance into phylogenetically independent comparisons (technically, linear contrasts) between related taxa. Both methods are derived from Felsenstein (1985). The first (Purvis & Rambaut 1995) compares values of a chosen trait – say, plant height – between one or more native species and one or more related aliens. As many as possible phylogenetically independent contrasts are computed. Under the null hypothesis that native and alien taxa are the same height, we should expect the aliens to be taller than natives in about half of the contrasts: the sign test is used to assess the significance of departures from this expectation. A second procedure (Pagel 1992) is used to test hypotheses involving more than two variables. This computes contrasts in each chosen variable at every node in the phylogeny; multiple regression through the origin is then used to test significance. We used the CAIC package (Purvis & Rambaut 1995) for both methods.

The methods we use require an estimate of phylogeny for all 2684 species in the dataset. Interfamilial relationships were taken from the strict consensus trees in Chase *et al.* (1993), with lower-level relationships according to Stace (1991). Families missing from Chase *et al.* were placed according to Stace (1991). Additionally, some families are paraphyletic or polyphyletic according to Chase *et al.'s* trees: where necessary, we used subfamilial divisions from Clapham, *et al.* (1962) to place species in our data set with their closest relatives in Chase *et al.* We set all branches in the phylogeny to the same length, transforming contrasts whenever they showed marked heterogeneity of variance (Harvey and Pagel 1991).

2.4 Statistics

Statistical analysis was confined to sign tests of the phylogenetically controlled contrasts, based on the null hypothesis of equal representation of positive and negative contrasts between the traits of native and alien plants. In cases where the value of the contrasts showed a trend with the trait score, a contingency table was constructed using the contrasts from the 12 largest and 12 smallest values of the trait score, with significance assessed at the 1% level ($\chi^2 > 6.64$) for a 2×2 table (positive and negative contrasts versus low and high trait scores).

3 Results

3.1 Numbers of natives and aliens

The numbers of alien and native taxa in Kent (1992) are shown in Table 3.1, along with estimates of the numbers of species of plants introduced intentionally and unintentionally to the British Isles (see Crawley 1997 for details). Of the 1169 naturalized alien species, about 70 have become sufficiently widespread in seminatural habitats that a visiting botanist might mistake them for natives. Only about 15 alien species are regarded as problem plants, but even amongst these species, there is far from unanimous agreement as to their pest status (Williamson 1993). Thus, somewhere between 0.5% and 5% of introduced species have become naturalized, and about 6% of naturalized species behave like natives. By even the most generous estimate, well under 0.1% of introduced species have become pests (cf. Williamson 1993).

3.2 Taxonomy

The aliens and natives were not drawn from species pools with similar taxonomic compositions. Many alien species belong to one of the 40 families

Table 3.1. *The numbers of native and alien species in the British flora with estimates of the numbers of alien species introduced into the British Isles for various purposes.*

(The largest number of introductions was of herbarium specimens; Kew gardens has over 7 million herbarium sheets representing about 80% of the total world flora. About 65 000 named taxa are currently sold for horticulture in Britain (Philip & Lord, 1995), of which about 14 000 represent distinct species grown out of doors. The casual flora runs to more than 6000 species.)

World			250 000
Brought to Britain	unintentionally	20 000	
	intentionally	200 000	
			220 000
Grown in Britain	botanic gardens	25 000	
	commercial horticulture	14 000	
	non-cultivated aliens	6000	
			26 000
British flora	native	1515	
	naturalized	1169	
			2684
Widely naturalized aliens			68
Problem plants			15

not represented in the native British flora, and many families have disproportionately many alien representatives (e.g. Pinaceae, Iridaceae; Kent 1992). It is noteworthy that five of the top 12 overrepresented families are phanerophytes (trees and shrubs) and that four of the top 12 are geophytes (bulbs and other plants that perennate underground). Likewise, there are many British plant families which have no alien representatives (including some large families such as Orchidaceae and Potamogetonaceae) and numerous families that have disproportionately few alien representatives (e.g. Cyperaceae, Juncaceae). The families with no alien members contain many ferns and water plants, two groups with particularly effective long-distance dispersal. The under-represented families contain a preponderance of graminoid monocots like sedges and rushes, more ferns and parasitic plants. The relative paucity of alien Caryophyllaceae is a puzzle.

This non-random species representation makes a phylogenetically controlled approach essential, otherwise counts of species (e.g. chi-squared analysis of contingency tables) might show nothing more than the idiosyncrasies of uneven representation of different plant families.

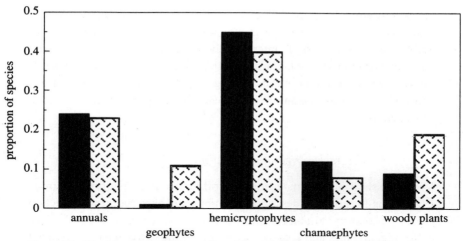

Figure 3.1. The frequency of different Raunkiaer life forms for natives (solid) and aliens (light stippling). The highly significant preponderance of trees and geophytes amongst the aliens is an artefact of the unequal representation of plant families and the predilection of gardeners for ornamental trees and bulbs.

3.3 Life forms of natives and aliens

The frequency distributions of Raunkiaer's life forms for the native and alien species are shown in Figure 3.1. There is a highly significant difference in the proportional representation of natives and aliens ($\chi^2 = 27.0$, d.f. = 4, $P < 0.001$) when species are treated as independent data points; woody plants and geophytes are strongly overrepresented amongst the aliens. As we have seen, however, there is a highly significant bias in the taxonomic representation of different alien plant families, and this is determined largely by the horticultural tastes of British gardeners rather than by ecological performance (i.e. gardeners like trees and bulbs). It is unwise, therefore, to interpret the contingency table as showing that different life forms are more or less likely to become established as aliens in the absence of any control over the rate of introduction of species. Phylogenetically controlled analysis of the life-form data is discussed below.

3.4 Mode of introduction

Most naturalized alien plant species were intentionally introduced. Much the largest category is made up by escaped garden plants, but plantation trees and woody crop plants are well represented. The majority of unintentionally introduced plants are casuals, although several important annual weeds of arable agriculture were unintentionally introduced, probably as contaminants of imported seed. The notion that the most important alien plants arrive as seeds stuck to bootlaces or caught up in trouser turn-ups is wrong. Most of our pernicious plant invaders were originally introduced as garden ornamentals.

3.5 Geographic origin of British aliens

Most British aliens originate from central and southern Europe (Crawley 1997). A log-linear model describing the number of alien species as a function of four explanatory variables (latitude, area of source country, size of the flora in the original range and great circle distance to the source region) contained significant terms (in order of importance) for latitude (positive), distance (negative) and log(area) (positive). There was no effect of the size of the local flora, presumably because this is well predicted by latitude and log(area). Significant outliers from this model were as follows: Europe, South America and New Zealand had more British aliens than predicted; North America, Turkey and the Middle East, and Mediterranean Europe had fewer British aliens than predicted. The general lack of frost hardiness amongst tropical plant species means that few, if any, tropical species are naturalized in the British Isles. This result draws attention to the asymmetry of exchange of aliens between different countries (e.g. North America received almost the entire European weed flora, but was the source of rather few European aliens; Crawley 1986, 1987).

3.6 Invaded habitats within the British Isles

Alien plant species are non-randomly distributed across habitats (Figure 3.2). Habitats rich in aliens tend to be created by humans or to be highly disturbed by human activities, and are often where there has been high to average cover of bare ground (e.g. waste ground, urban sites, railway lines, walls). Alien species richness is positively correlated with the rate of propagule introduction (e.g. proximity to gardens and allotments, proximity of seed sources like granaries, woollen mills, docks, roadsides and tanneries) and negatively correlated with the isolation of the habitat from urban influence. Thus, mountain tops and remote heathland areas are especially poor in alien species, and habitats like coastal sand dunes, which are rich in alien species

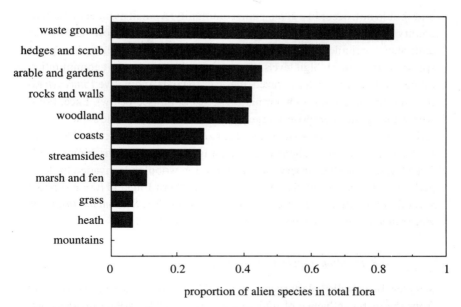

Figure 3.2. The habitats invaded by aliens in the British Isles, showing the fraction of the total flora of each habitat made up by alien species. Note the strong correlation with human disturbance and the negative relationship with isolation from human activities. These patterns are confounded by the fact that the rate of introduction of suitably pre-adapted alien species declines with isolation and with the difference between the ecological conditions of the target habitat and standard garden conditions (e.g. many more introduced plants are capable of growing in grassy waste ground than on upland cotton grass bogs).

where they occur close to towns (e.g. in Southern England and South Wales), are notably poor in alien plants in more remote areas (e.g. in the Outer Hebrides).

3.7 Abundance of alien plants

Data on the abundance (e.g. population density or biomass) of native plants are extremely scarce; data on the abundance of aliens are even rarer. Most data on plant abundance consist of subjective scores (e.g. Tansley's DAFOR system in which each plant species is allocated to one of the following categories: dominant, abundant, frequent, occasional or rare). There is also a pronounced sampling bias against alien plants because most of the data on plant abundance are gathered by phytosociologists, and these people exhibit an unusually high degree of chauvinism about which plant species should, and should not, occur in a given plant community. A small number of modern county floras contain semi-quantitative habitat lists that show no obvious 'anti-alien' prejudice; these data reinforce the impression that alien plant species seldom reach high levels of abundance in British plant communities. Some communities such as grazed, mesic grasslands and native *Pinus sylvestris* woodland contain no alien plant species at all (see below). Exceptions, where alien plants comprise a large part of the biomass, include some characteristic communities of waste ground (e.g. *Buddleja davidii* scrub or *Fallopia japonica* thicket where aliens often make up over 90% of the biomass; M. Crawley, unpublished observations), plantation woodlands invaded by *Rhododendron ponticum* and chalky sea cliffs in southern England (these support several abundant species including *Centranthus ruber*, *Smyrnium olusatrum*, *Antirrhinum majus*, *Lycium* spp. and *Carpobrotus edulis*).

3.8 Isolation from human influence

Very few British alien plant species are established at any great distance away from the direct influence of human activities. Crawley (1997) divides alien species on the basis of the '100 m rule'. Field work by the first author (M. J. Crawley, unpublished results) undertaken throughout the British Isles shows that the great majority of naturalized alien plant species are seldom found more than about 10 m from buildings, gardens, walls, waste ground, roadsides or railway lines. Only a handful of species is ever found more than 1000 m away from human disturbance. Of course, it is true that only a tiny fraction of the land area of the British Isles lies more than 1 km from a road or railway (rural roads tend to be about 1 km apart in lowland Britain). Furthermore, isolated areas contain an extremely biased sample of seminatural habitats (mostly heathland, mountain-top, streamside or saltmarsh). The '100 m rule' defines an approximate threshold that separates the small number of thoroughly naturalized aliens that appear not to need direct human intervention for their persistence (e.g. *Mimulus* hybrids, *Epilobium brunnescens*, *Oenothera glazioviana*) from the much larger number of species that appear to be incapable of regeneration in undisturbed seminatural vegetation.

3.9 Distribution of alien plants within the British Isles

There is a clear trend of declining alien species richness within the British Isles from a maximum in southern England, declining northwards and westwards

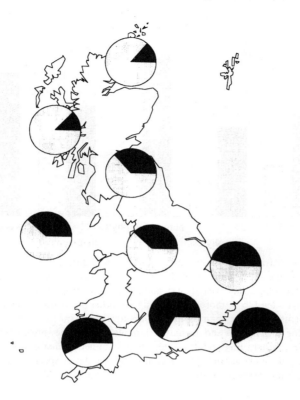

Figure 3.3. The geographic distribution of alien plant species within the British Isles. The pie diagrams show the proportion of the local flora made up of alien species (dark stippling). There is a pronounced gradient from 67% alien species in southern England to 13% in north-west Scotland. These trends reflect climatic differences (e.g. many frost-sensitive alien species are restricted to the south west of England, Isles of Scilly and the Channel Islands), habitat heterogeneity (there are more habitats per 10 km square in southern England than in northern Scotland) and differences in the rate of propagule introduction (this is probably roughly proportional to human population density, which shows a strong south-east to north-west gradient).

to a minimum in north-western Scotland (Figure 3.3). Local alien hotspots include the almost frost-free Isles of Scilly off the south-west coast of Cornwall (boosted by escapes of South African and New Zealand plants from the celebrated gardens at Tresco Abbey) and central London (possibly as a result of the warmer climate afforded by the urban 'heat island'). These patterns are consistent with the hypothesis that the most important factor determining the number of alien plant species in a local area is the rate of introduction of propagules by people.

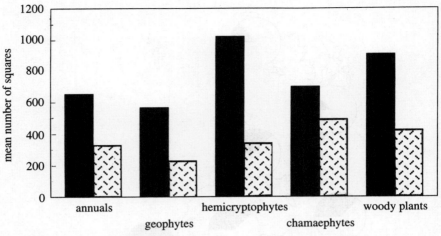

Figure 3.4. The distribution of native and alien species of different life forms, estimated as the average number of 10 km squares occupied per species. There are intriguing patterns for both natives (solid) and aliens (light stippling). Native hemicryptophytes tend to be more widespread than other life forms, while the most widespread aliens tend to be chamaephytes. Geophytes have the most restricted distributions for both natives and aliens, presumably because they are commonest on warmer soils in southern England. It is interesting that alien hemicryptophytes are proportionately the most restricted when compared with native species, suggesting relatively higher native species saturation than in other life forms like trees.

Some alien plant species, such as *Acer pseudoplatanus*, are virtually ubiquitous in their distributions within the British Isles, but the average range of alien plants (as determined by the number of 10 km squares from which they are recorded) is significantly more restricted than the average for natives (Crawley 1997). The most widespread alien, *Matricaria discoidea*, is the 33rd most widespread British species, but the 10th most abundant alien, *Elodea canadensis*, is only the 311th most widespread species.

There are intriguing differences in the spatial distributions of the different life forms (average number of 10 km squares occupied), and natives and aliens show different patterns. For natives, hemicryptophytes have the widest distributions, but alien woody plants and chamaephytes have wider distributions than alien hemicryptophytes. Geophytes have the most restricted distributions for both natives and aliens (Figure 3.4), largely because they are restricted to the warmer, base-rich soils of southern England.

Table 3.2. *Phylogenetically controlled contrasts for a variety of traits, comparing native and alien vascular plant species in the British Isles*

Alien plant species were taller and had larger seeds than the natives. Native plants had more extensive geographical distributions within the British Isles. Seed dispersal syndromes were different; more of the natives were water- or wind-dispersed; more of the aliens were dispersed directly by people or had explosive fruits. Species = the number of species for which data on the trait were available. Contrasts = the number of independent contrasts between natives and aliens in the available data. Signs = number of contrasts in which the native trait was greater than the alien trait/the number of contrasts in which the alien was greater than the native. When there are ties (native = alien), the sum of the positive and negative signs is less than the number of contrasts. Significance = 2-tailed probability of an equal or more extreme distribution of signs (binomial test).

Trait	Species	Contrasts	Signs	Significance
Plant height	1562	134	49/80	0.01
Seed weight	782	58	21/37	0.05
Distribution	1538	115	97/18	0
Dispersal	1669	143	44/25	0.03
Life form	1505	143	43/44	n.s.
Seedling growth rate	127	6	5/1	n.s.
Start flowering	1562	122	45/60	n.s.
End flowering	1556	120	55/52	n.s.
Pollination	1508	119	19/25	n.s.
Altitude	918	36	21/15	n.s.

3.11 Ecological traits of alien plants

Several traits showed no significant differences between native and alien species, but a few traits were strongly and consistently associated with aliens in phylogenetically independent contrasts (Table 3.2). The most clear-cut difference was in plant height; alien plants were taller than their native counterparts in 80 of 134 contrasts ($P < 0.01$). Given that plant height and seed size are positively correlated (Rees, this volume), it is not surprising that aliens tended to have larger seeds than their native counterparts. Seed bank dynamics of natives and aliens were significantly different; aliens were less likely to show no seed dormancy and more likely to show protracted (> 20 year) seed dormancy. There was a significant but small tendency for alien plant species to flower relatively early or relatively late in the year than native species, and to be more likely to be pollinated by insects. Within the category of aliens, date of introduction was significantly associated with geographic distribution (not surprisingly, recent introductions had more restricted ranges than did long-established aliens) and more recent introductions were likely to be smaller (and less likely to be trees) than early introductions.

Perhaps the most intriguing pattern to emerge from the phylogenetically independent contrasts has to do with Raunkiaer's life form. Ranking species in the sequence of increasing size and longevity (therophyte, geophyte, hemicryptophyte, chamaephyte and phanerophyte) showed no overall distinction between natives and aliens (compare with Figure 3.1 and with Williamson & Fitter 1996), largely because most of the contrasts are between species with the same life form (e.g. all Pinaceae and all Fagaceae are phanerophytes). Nevertheless, there was a significant surfeit of positive contrasts (aliens bigger and longer lived) amongst the long-lived species and a significant surfeit of negative contrasts (aliens smaller and shorter lived) amongst the short-lived species ($\chi^2 = 6.18$; d.f. = 1).

4 Discussion

Given that we had no *a priori* reason to expect that there would be any trait differences between native and alien plants (because all plants must possess traits which allow them to increase when rare), it is extremely interesting to find such a consistent set of relationships in data on the British flora. It looks as if aliens need to 'try harder'; they have bigger seeds and they are taller than comparable native species. These traits are positively correlated, as Rees (this issue) shows. There is evidence that native British communities are not saturated in species and that some life forms (e.g. hemicryptophytes) are more saturated than others (e.g. trees). Vacant niches do exist, but as we shall see, some aspects of the concept of vacant niche are relatively more clear-cut than others. There is no doubt that much clearer patterns will emerge in future, when detailed attention is focused on particular taxa in specified habitats with comparable disturbance regimes.

The first point that needs to be emphasised is that most alien plants were not introduced by accident (the 'seeds in trouser turn-ups' model). The vast preponderance of intentional over unintentional introductions amongst the British alien flora matches the experience from other well-studied floras. In South Africa, for example, government legislation classifies alien plants into two categories of severity: of the 10 species of 'declared invaders' which represent a serious threat to native ecosystems, all are woody and all were intentionally introduced. Of the 138 species of 'declared weeds', 82% were intentionally introduced, and amongst the 15 unintentionally introduced species, all were herbaceous, and most were pasture weeds (Henderson 1995).

This last point draws attention to a paradox concerning British grasslands. Grazed, mesic grasslands in Britain support no alien plant species at all, and yet these same grasslands were the source of many of the most pestilential pasture weeds introduced into other parts of the world (*Hypericum perforatum, Cirsium arvense, Senecio jacobaeae, Pilosella officinarum, Plantago lanceolata, Rumex acetosella, Ulex europaeus, Cytisus scoparius* and *Hypochoeris radicata* in Australia, South Africa, Pacific islands and the Americas). The probable cause of this asymmetry is the long association between Old World grasslands, people and grazing ungulates, and the importance of introduced domestic livestock as primary agents of disturbance in New World habitats, paving the way for invasion by pre-adapted, grazing tolerant, Old World pasture species when these were introduced in hay and seed mixtures imported by the early European settlers.

A model for the geographic origin of British aliens highlights the importance of proximity and climatic matching. The number of species originating from different regions declines monotonically with increasing difference in latitude (lack of frost-hardiness in tropical and subtropical species is the most obvious cause of this pattern). The strong negative relationship between the number of alien species and distance to the source country (e.g. Europe is the principal source of British aliens) is not a general phenomenon, because Europe is also the principal source of alien plants in distant locations like Australia, New Zealand and South Africa.

Although the British flora is better known than any other flora in the world, there are still a number of potentially serious biases that could influence our interpretation of the comparative ecology of alien plant species in Britain. Knowledge about the abundance of aliens and their ecological function in seminatural habitats is dogged by the tendency of botanists to avoid communities that contain aliens, or to locate their study quadrats in alien-free parts of invaded communities. Worse still is the lack of information on the autecology and ecophysiology of alien plants in sources like the ecological flora database (Fitter & Peat 1994).

Despite these difficulties, a few clear patterns have emerged. In phylogenetically independent contrasts, British aliens were larger than their native counterparts, had larger seeds, were more likely to flower very early or very late in year, had long-lived seed banks and were more likely to be pollinated by insects. Some of these traits are internally consistent (e.g. plant height and seed weight are positively associated), but others are not (e.g. large seeds and long seed dormancy are negatively associated in most between-species comparisons; Rees, this issue). The taking these observations together suggests that there are at least two characteristic groups of aliens: those species that are 'more K-strategist' than native K-strategists; and those that are 'more r-strategist' than native r-strategists. Thus, the alien flora contains an abundance of woody and thicket forming species that are capable of excluding native vegetation (*Rhododendron ponticum*, *Symphoricarpos albus*, *Fallopia japonica*) and a group of small, rapidly maturing, long-flowering species that soon succumb to interspecific competition during secondary succession (*Epilobium ciliatum*, *Veronica persica*, *Senecio squalidus*).

To the extent that the niches of the resident natives define the niche space occupied by an entire community, it is clear that there will always be vacant niches for alien invaders at 'both ends' of any niche axis. For every ordination of the resident native species (e.g. along an r–K continuum based on time since last disturbance, or on a ranking from most to least nutrient demanding) there are (at least) two possibilities for invasion by an alien species. For example, on a successional niche axis, the alien could grow taller than the native dominant (e.g. alien trees like *Pinus* spp. or *Acacia* spp. overtop and outcompete the native dominant Proteaceae in South African fynbos; Cowling 1992). On an axis of nutrient supply rate, an alien species could invade if it was less demanding in its nutrient requirements than the most tolerant native species (i.e. it had a lower R^* *sensu* Tilman 1988). We can already see the potential for British invasions of this kind. Given only a modest degree of climatic warming, trees like *Robinia pseudoacacia*, *Ailanthus altissimus* and *Quercus cerris* might begin to replace native dominants, as they have in other parts of Europe (Pysek *et al.* 1995). Similarly, there will always be the possibility of replacement of native early successional species by alien r-strategists that do things more quickly, or disperse their seeds more widely (e.g. *Epilobium ciliatum*, *Conyza sumatrensis*, *Galinsoga parviflora*).

For non-extreme positions on the niche axis, the value of the concept of vacant niche is much less obvious. Does the likelihood of establishment of an alien species 'in between' two native species depend upon their degree of niche separation? Does establishment of an alien species necessarily lead to the greatest reduction in abundance of those native species that are closest to it on a niche axis? Which of the many possible niche axes are predicted to be the most important in explaining or predicting alien invasions?

Despite the importance of chance and timing in the establishment of alien plants (Crawley 1989), invasions are clearly not completely random events. There are several predictable patterns:

1. weeds in one country are likely to become weeds when introduced to another climatically matched country;
2. invasive crop plants in one country will be invasive in other countries at similar latitudes;
3. the rate of establishment of alien plants will be proportional to the frequency and intensity of disturbance of the habitat (e.g. by alien ungulates);
4. the higher the rate of introduction of propagules, and the greater the degree of matching of the ecological attributes of the source and target habitats, the greater the number of alien species is likely to be;
5. alien plants will grow bigger and have greater ecosystem-level impact than equivalent native plant species as a result of release from their specialist pathogens and herbivores.

On the other hand, there appears to be rather little that can be said about the native distribution and abundance of non-weedy plants and their likely performance as aliens. In the case of introduced insects used for the biological control of weeds, there was a positive correlation between the probability of establishment and their distribution and abundance as natives (Crawley 1986, 1987). No such pattern is apparent for British alien plants. In the genus *Impatiens*, for example, the three alien species show an inverse correlation between the extent of their native range and the extent of the alien range within Britain; the most widespread and abundant alien in Britain, *I. glandulifera*, has the most restricted native range.

Most of the differences between the findings of the present analysis and those of earlier studies can be put down to the lack of any attempt at phylogenetic control in works reported in Drake *et al.* (1989), Cowling (1992) and Pysek *et al.* (1995). Their contrary opinions about the importance of life history, seed size and plant height can all be attributed to the problems associated with using chi-squared contingency tables to analyse species count data (Crawley 1993), when there is conspicuously unequal taxonomic representation of natives and aliens (Table 3.1). Our analyses assume that our estimate of phylogeny is correct. If it contains non-monophyletic groups, Type I error rates will be elevated. If not all true sister-taxon relationships are shown in the available phylogeny, Type II rates will rise. In the absence of true branch length information, both will rise. These caveats should be borne in mind when considering any of our findings that are near the threshold of significance or based on few contrasts. However, they should not be taken as reasons in favour of analyses treating species as independent; our estimate of phylogeny, however faulty, reflects reality much better than the assumption implicit in such a treatment (namely, that all species radiated instantaneously from a common ancestor). Indeed, simulations show that such cross-species analyses often have wildly elevated Type I error rates, far in excess of any rate found with independent contrasts, even when its assumptions are violated (Grafen 1989; Purvis, *et al.* 1994).

Finally, the appearance of similar-looking, 'reconstructed' plant communities made up almost entirely of alien species speaks of the existence of the kind of assembly rules that have proved to be exasperatingly difficult to detect in native vegetation (Wilson 1995). To judge by the few examples studied to date (e.g. mesic grasslands in Hawaii, urban scrublands in Britain) the key parameters are the resource supply rate (this determines the identity of the dominant plant species) and the life history, size and longevity of the dominant alien species (e.g. *Ulex europaeus* in Hawaii, *Buddleja davidii* in Britain). These alien plant communities represent a rich source of natural experiments on community assembly.

References

Chase, M. W., Soltis, D. E., Olmstead, R. G. *et al.* (1993). Phylogenetics of seed plants: an analysis of nucleotide sequences from the plastic gene *rbc*L. *Annals of the Missouri Botanical Garden* **80**, 528–580.

Clapham, A. R., Tutin, T. G. & Warburg, E. F. (1962). *Flora of the British Isles.* Cambridge University Press.

Clement, E. J. & Foster, M. C. (1994). *Alien plants of the British Isles.* London: BSBI.

Cowling, R. M. (1992). *The ecology of fynbos: nutrients, fire and diversity.* Oxford University Press, Oxford.

Crawley, M. J. (1986). The population biology of invaders. *Philosophical Transactions of the Rural Society of London* **B 314**, 711–731.

Crawley, M. J. (1987). What makes a community invasible? In *Colonization, succession and stability* (ed. A. J. Gray, M. J. Crawley & P. J. Edwards), pp 429–453. Oxford: Blackwell Scientific Publications.

Crawley, M. J. (1989). Chance and timing in biological invasions. In *Biological invasions, a global perspective* (ed. J. A. Drake *et al.*) pp 407–423. Chichester: John Wiley.

Crawley, M. J. (1990). The population dynamics of plants. *Philosophical Transactions of the Royal Society of London* **B 330**, 125–140.

Crawley, M. J. (1993). *GLIM for ecologists.* Oxford: Blackwell Scientific Publications.

Crawley, M. J. (1996). *Plant Ecology*, 2nd edn. Oxford: Blackwell Scientific.

Crawley, M. J. (1997). *Aliens: the ecology of the non-indigenous flora of the British Isles.* Oxford University Press.

Drake, J. A. *et al.* (1989). (eds) *Biological invasions, a global perspective.* Chichester: John Wiley.

Felsenstein, J. (1985). Phylogenies and the comparative method. *American Naturalist* **125**, 1–15.

Fitter, A. H. & Peat, H. J. (1994). The ecological flora database. *Journal of Ecology* **82**, 415–425.

Grafen, A. (1989). The phylogenetic regression. *Philosophical Transactions of the Royal Society of London.* **B. 326**, 119–157.

Grime, J. P., Hodgson, J. G. & Hunt, R. (1988). *Comparative plant ecology: a functional approach to common British species.* London: Unwin Hyman.

Harvey, P. H. & Pagel, M. D. (1991). *The comparative method in evolutionary biology.* Oxford University Press.

Harvey, P. H., Read, A. F. & Nee, S. (1995). Why ecologists need to be phylogenetically challenged. *Journal of Ecology* **83**, 535–536.

Henderson, L. (1995). *Plant invaders of Southern Africa.* Pretoria: Plant Protection Institute.

Kent, D. H. (1992). *List of vascular plants of the British Isles.* London: BSBI.

Pagel, M. D. (1992). A method for the analysis of comparative data. *Journal of Theoretical Biology* **156**, 431–442.

Philip, C. & Lord, T. (1995). *The plant finder.* London: Royal Horticultural Society.

Purvis, A., Gittleman, J. L. and Luh, H.-K. (1994). Truth or consequences: effects of phylogenetic accuracy on two comparative methods. *Journal of Theoretical Biology* **167**, 293–300.

Purvis, A. & Rambaut, A. (1995). Comparative analysis by independent contrasts (CAIC): an Apple Macintosh application for analysing comparative data. *Computer Applications in the Biosciences* **11**, 247–251.

Pysek, P., Prach, K., Rejmanek, M. & Wade, M. (1995). *Plant invasions: general aspects and special problems.* Amsterdam: SPB Academic Publishing.

Rees, M. (1995). EC–PC comparative analyses? *Journal of Ecology* **83**, 891–893.

Stace, C. (1991). *New flora of the British Isles.* Cambridge University Press.

Tilman, D. (1988). *Plant strategies and the dynamics and structure of plant communities.* Princeton University Press.

Williamson, M. (1993). Invaders, weeds and the risk from genetically modified organisms. *Experientia* **49**, 219–224.

Williamson, M. & Fitter, A. (1996). The characters of successful invaders. *Biological Conservation.* (Submitted.)

Wilson, J. B. (1995). Testing for community structure: a Bayesian approach. *Folia Geobotanica et Phytotaxonomica* **30**, 462–469.

II • Reproductive traits

4 · The comparative biology of pollination and mating in flowering plants

Spencer C.H. Barrett, Lawrence D. Harder and Anne C. Worley

1 Introduction

Among life-history traits, reproductive characters that determine mating patterns are perhaps the most influential in governing macroevolution. This fundamental role arises because the mating system (who mates with whom and how often) governs the character of genetic transmission between generations and hence the behaviour of all genes in populations. Important evolutionary processes responsible for the diversification of plant lineages, including reproductive isolation and speciation modes, are closely linked to changes in mating patterns through their effects on the genetic structure and evolutionary dynamics of populations (Stebbins 1974). Indeed the diversity of many angiosperm families, such as the Orchidaceae (Dressler 1981) and Polemoniaceae (Grant & Grant 1965), has been directly attributed to the evolutionary flexibility of their reproductive systems.

Evolutionary responses to ecological conditions are evident in all aspects of angiosperm reproduction, from resource investment in reproductive versus vegetative function, through the structure and arrangement of flowers and their role in pollination, to mating patterns within populations. Hence examples of adaptation can be found in the relatively large reproductive effort of annuals compared with their perennial relatives (Primack 1979), the convergence in flower structure and colour of unrelated species with similar pollinators (Fægri & van der Pijl 1979) and the higher incidence of selfing among colonizing species than among related taxa occupying more stable habitats (Lloyd 1980). Because resource allocation, pollination and mating determine reproductive success in an integrated way, functional correlations between these components of reproduction should be commonplace. These functional linkages require that evolution involves coordinated changes to all aspects of reproduction in concert with life-history evolution. Unfortunately these linkages have often been overlooked because of the fragmentation of reproductive biology into subdisciplines specializing in different phases of reproduction.

The preceding perspective implies that between-species comparisons of resource allocation, pollination and mating should provide a rich source of insights into the ecology and evolution of plant reproduction. Although comparative analysis has a venerable tradition among reproductive botanists (e.g. pollination and dispersal syndromes), it was used relatively little during the 1970 and 1980s due to the dominance of species-level microevolutionary approaches that accompanied the growth of population biology. However, recognition of the utility of historical reconstruction and comparative biology for testing ecological and evolutionary hypotheses has recently awakened an interest in phylogenetics among evolutionary ecologists interested in plant reproduction and, as a result, studies increasingly consider a broader range of taxa with diverse life histories.

Reproduction is amenable to evolutionary analysis using both comparative and experimental approaches for various reasons. Floral traits are often well documented because of their taxonomic importance, hence surveys of floras and monographs often provide valuable data. Plants display considerable inter- and intraspecific variation in floral traits that influence mating patterns, indicating the evolutionary lability of most reproductive characters. This variation increases the chances of detecting patterns in comparative data as well as providing opportunities for genetic and microevolutionary studies. Theoretical models of the evolution of reproductive systems often make specific predictions of both expected character associations and the likely order of establishment of traits in a phylogeny (e.g. the evolution of dioecy: Givnish 1980, but see Donoghue 1989). In some cases (e.g. the evolution of heterostyly; Charlesworth & Charlesworth 1979; Lloyd & Webb 1992) models differ concerning proposed evolutionary sequences. Phylogenetic reconstruction facilitates resolution of such conflicts.

In this chapter we illustrate how comparative approaches can aid studies of plant reproductive adaptations. We begin by clarifying our perspective on what constitutes an adaptation to avoid the confusion caused by recent disagreements about this concept between workers in phylogenetics and comparative biology, on one hand, and those concerned with mechanisms of selection in contemporary populations, on the other. We then investigate three current issues in plant reproductive biology using comparative approaches that differ in the availability of phylogenetic information. For each case we review the theoretical foundation upon which the problem rests and then ask whether comparative approaches provide insights not obtainable by microevolutionary enquiry. The first problem concerns the allocation of sexual resources as it pertains to the ratio of pollen and ovules produced by flowers. The last two issues relate to mating patterns, beginning with whether particular aspects of life history are associated with selfing and outcrossing, and then proceeding to a specific application of phylogenetic reconstruction to an analysis of the evolutionary assembly of a complex reproductive adaptation governing the mating system. Rather than present detailed results and analyses for these problems, we emphasize unanswered questions that would benefit from research conducted from a historical perspective and identify difficulties inherent in implementing and interpreting comparative analyses.

2 Adaptation, ecology, evolution and comparative analyses

Comparative analyses are commonly used to identify and interpret adaptations. The perceived success of this exercise often depends on one's understanding of what adaptations are and how they arise and are maintained. Although we do not intend to join the debate about the 'true' meaning of adaptation (reviewed by Reeve & Sherman 1993), our discussion of reproductive adaptations in plants is less susceptible to misinterpretion if we clarify our use of this concept.

Most generally, an adaptation refers to a trait (or complex of traits) that confers functional advantage over alternative traits in a given environment. Association of phenotype with function recognizes that adaptations have purposes which contribute to an organism's fitness. Consequently, covariation between two characters among taxa informs us about adaptation only to the extent that we understand their functional linkage, rather than, for example, their developmental association. By including the environmental context in our definition we further identify that specific traits do not usually have intrinsic ecological value. For plant reproduction this is evident from the association of intraspecific geographical variation in flower design and mating system with variation in pollinator diversity and abundance (e.g. Grant & Grant 1965).

As a result of their functional role, adaptations are maintained through natural selection, thereby achieving evolutionary relevance. Typically, adaptations originate through natural selection, although this need not be universal. For example, if evolutionary divergence involves Wright's (1977, chapter 13) shifting balance, genetic drift could theoretically produce novel traits which subsequently gain status as adaptations if they are more beneficial than alternative traits. Because the demographic conditions prevalent during the

Table 4.1. *Evolutionary interpretations of traits based on comparisons of species*

	Related species			Unrelated species	
	Similar functions	Different functions	Similar functions	Different functions
Similar traits	1a parallel evolution 1b phylogenetic niche conservatism 1c phylogenetic 'inertia' or 'constraint'	2 exaptation for the species with the novel function	3 phenotypic convergence	4 trait similarity uninformative
Different traits	5a alternative phenotypic solutions 5b trait *differences* non-functional	6 adaptive trait divergence	7 functional convergence	8 interpretation uncertain, possibly adaptation

origins of traits are unknown and unknowable, assertion that any particular adaptation arose solely through natural selection, let alone that selection favoured a particular function, cannot be justified.

In the context of comparative biology, adaptations can be recognized in several guises, which are characterized by the extent of phenotypic and functional similarity between taxa (Table 4.1). Adaptation is readily recognized when unrelated species perform similar functions, regardless of whether similar or different traits are involved (interpretations 3 and 7 in Table 4.1). Similarly, trait divergence in related species is typically accepted as evidence of adaptation (interpretations 5 and 6), especially when it is associated with functional differences. On the other hand, comparisons of traits that perform different functions in unrelated species provide little insight into adaptation (interpretations 4 and 8) in isolation from directed studies of the fitness consequences of phenotype–function associations within each species.

Unlike the preceding cases, trait similarity in related species provokes considerable debate about the interpretation and nature of evidence for adaptation (see Reeve & Sherman 1993). We recognize function as an integral component of an adaptation, which is not secondary to phenotype, so that our definition of adaptation differs from that of Gould & Vrba (1982) by encompassing exaptation (interpretation 2). More fundamental dispute arises in the context of related species that use similar traits to perform similar functions (interpretation 1). Functional and phenotypic similarity between closely related species arises because common ancestry bequeaths related species with similar phenotypes and ecological niches. Resemblance of related species could persist if characters are resistant to selection, even though the species occupy different environments ('phylogenetic inertia': uniovulate florets in Asteraceae may provide one example). Probably more commonly, niche similarity leads to similar selection, so that related species continue to resemble each other because either environmental stasis results in equivalent stabilizing selection so that mean phenotypes remain unchanged, or similar environmental changes promote parallel evolution. To the extent that selection maintains the competitive advantage of the fit between phenotype and function, similarity of related species is consistent with our conception of adaptation.

The perceived importance of resemblance of related species depends on the perspective of the investigator. Ecologists are particularly interested in the association between form and function, whether it arises from directional or stabilizing selection, so both interpretations 1a and 1b (but not 1c) merit attention. In contrast, for some evolutionary biologists, similarity of related species due to analogous stabilizing selection is less interesting because it lacks the essence of evolution, namely change (e.g. Brooks & McLennan 1991; Harvey & Pagel 1991). Consequently, definitions of adaptation proposed by several evolutionary biologists (reviewed by Reeve & Sherman 1993) explicitly exclude interpretations 1*a* and 1*b* (in addition to interpretation 1*c*) from consideration. Furthermore, this rejection of phenotypic similarity among related species as an interesting phenomenon, even if that similarity is maintained by stabilizing selection or parallel evolution, is a fundamental feature of recently developed techniques for analysing comparative data (reviewed by Harvey & Pagel 1991). It is not surprising, therefore, that disagreements arise between proponents of these perspectives concerning the interpretation of comparative data (e.g. Harvey *et* al. 1995; Westoby *et* al. 1995), given their somewhat different conceptions of adaptation.

Regardless of one's perspective on adaptation, analysis of comparative data can present a technical problem. Statistical tests (parametric or nonparametric) that compare a test statistic to a prescribed probability distribution (e.g. F or χ^2) rely on the assumption of independent observations (i.e. $P[X = x_i \ \cap \ X = x_j] = P[X = x_i]P[X = x_j]$, where $P[X = x_i]$ and $P[X = x_j]$ are the probabilities that the values of some trait (X) for observations i and j are x_i and x_j, respectively, and $P[X = x_i \cap X = x_j]$ is the joint probability of observations i and j having these trait values). Depending on the question of interest, related species may not be independent observations from the statistical population of species phenotypes due to resemblance inherited from their common ancestor (Felsenstein 1985), leading to rejection of the statistical null hypothesis more often than expected (i.e. too many Type I errors). Evaluation of the independence assumption is often not straightforward in comparative analyses. In particular, phylogenetic relatedness of phenotypically similar taxa is not sufficient information to judge independence: the question under consideration provides the context for this assessment. As an analogy, consider whether identical twins who were separated at birth constitute independent observations. Clearly they do not if the trait of interest is eye colour, whereas they probably are independent subjects with respect to the current values of their bank accounts, even if those values are very similar. Similarly, the inclusion of related species in comparative analyses may or may not cause independence problems, depending on the topic and the relevant traits (see also Westoby *et* al. 1995). Unfortunately, lack of independence cannot be assessed by comparing results from statistical analyses that incorporate phylogeny (reviewed by Harvey & Pagel 1991) with those that do not, because current phylogenetically sensitive techniques define independence based on the phylogeny of taxa, rather than on the histories of the traits of interest. As the twin analogy indicated, trait history is the relevant feature in assessing independence of observations.

3 Floral allocation strategies

Plant reproduction involves hierarchical allocation of resources among inflorescences, between flowers and within flowers. For species with hermaphrodite flowers, allocation within flowers includes relative investment in pollen and ovules. To the extent that intrafloral allocation to gametes is limited and independent of allocation among flowers and inflorescences, investment in pollen should vary inversely with investment in ovules and seeds. Charnov (1982) proposed that the optimal resolution of this trade-off occurred when investment in pollen and ovules produced equivalent marginal fitness returns through both sex roles. The specific optimum for a given species depends on the relation between allocation in a particular gamete type and the realized genetic contributions (sex-specific gain curves). The shapes of these gain curves depend on many aspects of reproduction, including resource availability, pollinator attraction, pattern of pollen export and import, post-pollination processes, inbreeding depression and seed dispersal. As a consequence, species with different reproductive ecologies should produce pollen and ovules in correspondingly different ratios.

Unlike many aspects of plant reproduction, pollen–ovule ratios have been the subject of several comparative analyses. On the basis of an extensive survey, Cruden (1977) demonstrated that outcrossing species had higher pollen–ovule ratios than predominant selfers. He explained this pattern in terms of pollination efficiency, arguing that pollination is easily realized for selfers and so requires production of few pollen grains per ovule, whereas outcrossing involves the uncertainties that accompany reliance on pollen vectors and necessitates greater pollen production. This perspective was criticized (Charnov 1982) as equating a plant's reproductive success with its seed production (i.e. female function). Although justified, this criticism diverted attention from the importance of pollination and post-pollination processes in determining the shapes of male and female gain curves. Consequently, we believe many ecological and evolutionary influences on sex allocation await intraspecific and comparative analysis.

Investment in pollen involves both the number and size of grains. When the proportion of resources allocated to male function (r) is fixed, the number of pollen grains produced (P) must vary inversely with the resources invested

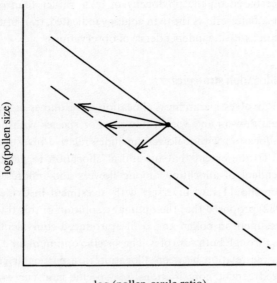

Figure 4.1. Expected relations between log(pollen size) and log(pollen–ovule ratio) for selfing and outcrossing species with equivalent seed size. The arrows indicate changes in pollen size that could accompany evolution of autogamy from an outcrossing ancestor.

per grain (S), so that $P \propto r/S$. Incorporating this trade-off with a similar relation for ovule and seed production, Charnov (1982) proposed that, for species with equivalent seed size, pollen size should vary with pollen–ovule ratio (P/O) according to

$$\log (S) = \log\left(\frac{r}{1-r}\right) - \log (P/O)$$

Because outcrossers should invest relatively more in male function than selfers (Charnov 1982), $r/(1 - r)$ will be greater for outcrossers than for selfers, resulting in an elevated trade-off relation between pollen size and pollen-ovule ratio for outcrossers (Figure 4.1). Given these relations, how should pollen size differ between selfers and outcrossers? As Figure 4.1 illustrates, both positive and negative differences are possible, so that the answer cannot be found simply by recognizing the size–number relation between pollen size and number and the greater investment in male function by outcrossers.

We must also consider functional relations between pollen size and mating system.

At least two features of reproduction suggest that outcrossers might produce larger pollen than selfers. First, large pollen is often associated with long styles (e.g. Williams & Rouse 1990; Kirk 1992), presumably because more resources are required for successful pollen-tube growth and fertilization. Autogamous species typically have small flowers with shorter styles, and so should produce smaller pollen. Second, pollen of outcrossers can be subject to intermale competition and female mate choice during pollen-tube germination and growth, so that large pollen may convey a competitive advantage, whereas the pollen of selfers need not be as competitive, so selfers should produce smaller, more economical, pollen. Hence details of reproductive biology lead to expectations for associations between pollen–ovule ratio, pollen size and mating system that are not apparent from sex-allocation theory, as currently developed.

We examined these expectations for selected selfing and outcrossing species in the Polemoniaceae. This family of approximately 320 species exhibits considerable variation in life history, growth form, pollination and mating (Grant 1959; Grant & Grant 1965: also see section §4). Opportunities for comparative analyses within this family improved recently with the publication of several phylogenetic studies based on DNA sequences of 20–59 species (reviewed by Johnson *et al.* 1996). On the basis of these phylogenetic trees, we selected six pairs of closely related congeners in which one member was autogamous and the other outcrossing (according to Grant & Grant 1965: *Collomia linearis–C. grandiflora, C. tinctoria–C. rawsoniana, Eriastrum wilcoxii–E. densifolium, Gilia leptomeria–G. rigidula, Langloisia setosissima–L. punctata, Navarretia divaricata–N. intertexta*). Comparison of these pairs indicates that outcrossers have higher pollen–ovule ratios than selfers (mean $\pm$ SE difference in $P/O = 204 \pm 59.60$, paired t-test, $t_5 = 3.42$, $P < 0.025$; data from Plitmann & Levin 1990), and outcrossers also produced larger pollen than selfers (mean $\pm$ SE difference in log[pollen volume, μm^3] $= 0.63 \pm 0.20$, $t_5 = 3.21$, $P < 0.025$: data from Stuchlik 1967). Interestingly, this pattern also occurs intraspecifically between selfing cleistogamous flowers and outcrossing chasmogamous flower of *Collomia grandiflora* (Lord & Eckard 1984). Despite the small sample, this result implicates pollen size, as well as pollen–ovule ratio and secondary reproductive characters (e.g. corolla size), in mating-system evolution. This conclusion exposes the need for increased attention to interactions between sex allocation, pollination and post-pollination processes, in general, and the functional significance of pollen size, in particular.

4 Mating systems and life history

Mating systems of seed plants range from obligate outcrossing, through simultaneous outcrossing and selfing (mixed mating), to predominant selfing (autogamy). This diversity is distributed non-randomly with respect to taxonomy and life history (Barrett & Eckert 1990), implying that full understanding of mating-system evolution requires consideration of both phylogeny and ecology. Surprisingly few comparative data have been collected on the ecological and life history correlates of outcrossing and selfing, despite considerable natural history information from groups with known mating systems (reviewed in Lloyd 1980). Moreover, few studies have attempted to reconstruct the phylogeny of outcrossing and selfing using formal cladistic techniques (although see Kohn *et al.* 1996; Schoen and co-workers unpublished).

Theoretical models concerning the relative frequencies of self- and cross-fertilization (reviewed in Uyenoyama *et al.* 1993) have primarily focused on genetic issues, including the transmission characteristics of genes modifying the reproductive system and the relative fitness of selfed and outcrossed individuals. Few models have explicitly considered joint effects of life history and mating system, even though natural history observations suggest associations between the woody growth form and outcrossing, and the annual habit and selfing (reviewed in Lloyd 1980). To assess whether interspecific variation in mating patterns correlates with life history, Barrett & Eckert (1990) surveyed quantitative estimates of selfing rate obtained using genetic markers and classified 129 species according to growth form (annual, herbaceous perennial or woody perennial). They found relatively more outcrossing among woody perennials than among annual and perennial herbs, but noted that the sampled species might not be statistically independent because, for example, the woody perennials included many related species (especially pines and eucalypts).

The recently published *rbcL* phylogeny of seed plants (Chase *et al.* 1993) provides an opportunity to reassess this association. We first surveyed recent literature to update Barrett & Eckert's (1990) dataset, resulting in a sample of 217 species from 43 families. Because the Chase phylogeny primarily depicts relationships between families and is based on different species from those sampled for selfing rates, we sought higher resolution phylogenies for five families (Asteraceae, Fabaceae, Onagraceae, Pinaceae, Poaceae) represented by several species in the selfing-rate data. Based on the inferred patterns of relatedness we constructed phylogenetically independent contrasts (Purvis & Rambaut 1995) between taxa with contrasting growth forms. These contrasts measure the difference in selfing rate between growth forms, which were then tested for consistent deviation from equality with signed-rank tests.

The results from this analysis (Figure 4.2) differed somewhat from those reported by Barrett & Eckert (1990). As with the earlier analysis, we found that woody perennials generally self less than herbs ($P < 0.0005$, $N = 18$). However, our results also reveal that annual herbs tend to self more than perennial herbs ($P < 0.0005$, $N = 20$), whereas Barrett & Eckert found no significant difference between these growth habits. One interpretation of the lack of accord in results for annual and perennial herbs between the two studies is that the earlier analysis suffered from inclusion of non-independent observations. Although this may be true, a second, perhaps more important, interpretation is that the analysis based on contrasts is more powerful because the matching of related species with alternate growth habits provides some control over diverse unmeasured ecological variables that also affect selfing rate. Unfortunately, the relative contribution of phylogeny and ecology as the prime influence on a species' selfing rate cannot be distinguished statistically with opportunistic surveys. A more convincing, though experimentally demanding, assessment requires two explicit contrasts: unrelated annual versus perennial herbs in the same environment to control for aspects of ecology other than those related to life cycle, and related annuals versus perennials from different environments to control for phylogeny.

Growth form incorporates several life history features that affect the advantages and disadvantages of different mating systems, including longevity, size and breeding frequency. These aspects of life history have important genetic and reproductive consequences. Long-lived species tend to have higher genetic loads resulting in strong inbreeding depression. This may constrain the evolution of selfing and explain why few woody plants are reported with high selfing rates (Figure 4.2a). In addition, because long-lived plants are often large and therefore susceptible to geitonogamous selfing, the resulting inbreeding depression may favour evolution of anti-selfing mechanisms (e.g. dioecy, self-incompatibility), which are common among woody plants and more prevalent among perennial than annual herbs. In the face of uncertain conditions during flowering, an annual life cycle is risky unless reproductive success is assured. Selfing provides such assurance, so the high incidence of predominant selfing among our sample of annuals (Figure 4.2c) is not surprising. These conditions suggest that life-history evolution prompted by altered ecological circumstances probably bears important consequences for mating-system evolution, and that the frequency of selfing is particularly responsive to changed conditions.

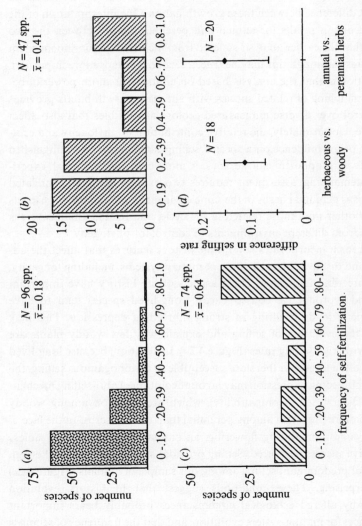

Figure 4.2. Relations between growth form and selfing rate for 217 species of seed plants. (a) Distribution of selfing rate for woody perennials; (b) distribution of selfing rate for herbaceous perennials; (c) distribution of selfing rate for annual species; (d) illustrates the mean ($\pm$ 95% confidence intervals) difference in selfing rate between herbaceous (annual and perennial) versus woody species and annual versus perennial herbs following phylogenetically independent contrasts (N = number of contrasts). See text for further details.

The evolutionary vagility of mating systems is evident in families with a significant number of herbaceous species, particularly those with a mixture of perennials and annuals (Stebbins 1974). For example, Grant & Grant (1965) noted that in the Polemoniaceae all autogamous species but one were annuals, although not all annuals were selfers, and they recognized that autogamy evolved repeatedly from outcrossing in annual species. With the recent publication of molecular phylogenies (see §3) the responsiveness of mating systems in the phlox family can be established more precisely by reconstructing the likely historical sequence and frequency of these changes. On the basis of *mat*K sequences for 77 taxa in this family plus two outgroups (L. A. Johnson, unpublished data) we identified 80 equally parsimonious phylogenetic trees with PAUP 3.1.1 (Swofford 1991; program options as in Johnson *et al.* 1996). Although the large number of trees suggests a poorly resolved phylogeny, the trees agreed in the overall historical pattern and disagreed primarily in the resolution of polytomies within four different clades (*Allophyllum*, *Navarretia* and the related *Gilia* sections *Gilia* and *Arachnion*). Onto these trees we mapped life cycle (annual or perennial), whether a taxon reproduced autogamously or not, and the incidence of self-incompatibility (data from Grant & Grant 1965, Plitmann 1994 and assorted floras) with MacClade 3.0 (Maddison & Maddison 1992). Only eight of the 80 trees depicted the combined evolution of *mat*K and mating system with equivalent parsimony: we selected one of these trees (Figure 4.3) to study mating-system evolution.

On the basis of the *mat*K phylogeny, the Polemoniaceae comprises a basal clade of tropical trees and shrubs, the monotypic, shrubby genus *Acanthogilia*, and a large, primarily herbaceous, temperate clade (Figure 4.3; see also Johnson *et al.* 1996). Within the temperate clade the annual life cycle evolved at least seven times and is a basal character for a large clade that includes *Allophyllum*, *Collomia*, *Navarretia* and *Gilia* sections *Gilia* and *Arachnion*. Reversion from an annual to perennial habit apparently occurred at least three times. Autogamy is restricted to the temperate clade and evolved independently from outbreeding at least 14 times, including nine within the *Allophyllum–Gilia* (*Arachnion*) clade, whereas there are no unequivocal instances of the opposite transition. Interestingly, self-incompatibility occurs in seven seemingly isolated terminal taxa, raising the possibility of multiple origins within the family (see Grant & Grant 1965, page 160).

The Polemoniaceae clearly exhibits evolutionary plasticity of reproductive mode and life history, implying that phylogenetic constraints do not limit opportunities when ecological conditions demand shifts in pollination and mating systems. Self-incompatibility in this family occurs in annual and perennial species with very different pollinators (including specialized flies, bumble bees, hummingbirds, hawkmoths, bats). Hence, there seem to be few limits to the development of self-incompatibility when plants consistently receive sufficient pollen that they can afford to reject self-pollen which would expose their offspring to inbreeding depression. Similarly, recurrent evolution of an annual life cycle, primarily among desert species, suggests considerable vagility of life cycle, at least among herbaceous species. The great diversity of the *Allophyllum–Gilia* (*Arachnion*) clade (Figure 4.3) also implies that annual life cycles persist in that lineage because such a life cycle represents a shared, ecologically appropriate solution, rather than a constrained life history. Indeed, the occasional reversion to a perennial life cycle within annual lineages illustrates that life cycle options remain open when they become more suitable for current ecological conditions. This is not to say that the course of evolution is unbounded. For example, the repeated origin of auto-gamy in lineages of annuals suggests that the reproductive assurance pro-vided by autogamy is a significant evolutionary influence once species adopt a high risk annual habit (Lloyd 1980). However, the generally terminal occurrence of autogamy in the phylogeny (Figure 4.3) and the absence of reversions to outbreeding imply that reliance on predominant selfing may be an expedient solution with a limited evolutionary future. In this group, at least, the evolution of selfing may be viewed as an 'evolutionary dead end' (Stebbins 1957).

Figure 4.3. Historical hypothesis of the evolution of annual life cycle and autogamy in Polemoniaceae. The phylogenetic tree is extracted from one of the eight most parsimonious trees based on *mat*K sequence analysis of 77 ingroup and two outgroup taxa (*Sarracenia purpurea*, Sarraceniaceae, *Fouquieria splendens* Fouquieriaceae: sequences provided by L. A. Johnson; consistency index = 0.664, retention index = 0.853), but only ingroup taxa for which the mating system is known are included here. The histories of mating-system and life-cycle transitions were reconstructed independently on the complete tree according to an equally weighted, unordered optimization scheme. Branch shading depicts the inferred sequence of changes in mating system (black branches denote at least partially outbreeding; open branches denote autogamy; hatched branches denote equivocal), whereas bars across branches indicate inferred transitions in life cycle (open bars denote perennial to annual; closed bars denote annual to perennial; stars adjacent to open bars indicate uncertain timing of a transition from perennial to annual).

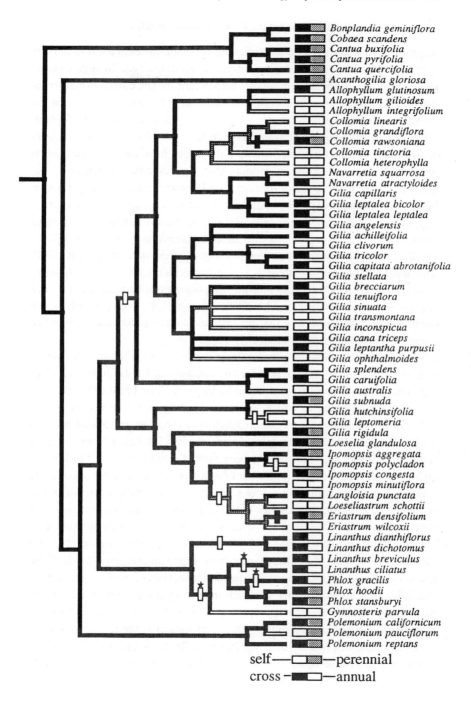

self——□▨——perennial

cross——■□——annual

5 The evolution of complex reproductive adaptations

Reproductive adaptations often comprise suites of correlated characters or syndromes that originate independently in different lineages through convergent selection. Comparative methods help determine the correlational structure of syndrome traits, independent of phylogeny; however, inference of the direction and temporal sequence of syndrome assembly requires explicit, optimised mapping of the characters of interest onto cladograms (Maddison & Maddison 1992). Our final example illustrates how this approach helps distinguish between selection models that predict different temporal sequences for the association of traits in a well known reproductive syndrome, heterostyly. A heterostylous population includes two (distyly) or three (tristyly) morphs that differ reciprocally in the relative positions of their male and female reproductive organs (reciprocal herkogamy) and various correlated ancillary polymorphisms of pollen and stigmas (reviewed in Barrett 1992). The floral morphs are usually self- and intramorph incompatible, with compatible matings resulting from intermorph pollinations between anthers and stigmas of equivalent height. This particular syndrome exemplifies how the evolution of pollination and mating systems is often functionally linked in complex reproductive adaptations.

Models of the evolution of heterostyly conflict as to whether heteromorphic incompatibility originates before (Charlesworth 1979; Charlesworth & Charlesworth 1979) or after reciprocal herkogamy (Lloyd & Webb 1992) and hence in their respective emphasis on anti-selfing versus improved cross-pollination as selective forces responsible for the evolutionary assembly of the polymorphism. Most heterostylous taxa possess incompatibility, so that self-compatibility has traditionally been viewed as a derived trait in heterostylous groups. However, the number of reported self-compatible species has increased in recent years (Barrett & Cruzan 1994), enabling inference of the sequence of evolutionary events within families containing heterostylous species that differ in the presence of incompatibility, such as Pontederiaceae.

Graham & Barrett (1995) and Kohn et al. (1996) reconstructed the phylogenetic history of Pontederiaceae from approximately two-thirds of the species in the family using partial sequences from the chloroplast genes rbcL and ndhF and chloroplast DNA restriction-site variation. Reproductive traits were mapped onto trees using optimization schemes that differed in the relative weights assigned to shifts in character states. On the basis of plausible weighting schemes (see below), self-incompatibility appears to have originated after the origin of floral trimorphism. This sequence supports Lloyd &

Webb's proposal that reciprocal herkogamy evolves initially to promote more proficient cross-pollination among individuals by more precisely matching pollen dispatch and receipt points on a pollinator's body. Incompatibility develops subsequently, either as a passive consequence of co-adaptation of each class of pollen to the style morph to which it is most proficiently transferred, or as an actively selected anti-selfing device. Several experimental results also support this evolutionary sequence (Barrett & Glover 1985; Kohn & Barrett 1992; Cruzan & Barrett 1996).

Although this concordance between phylogenetic and functional evidence appears satisfying, several issues concerning the historical reconstructions warrant consideration. Topological uncertainties in tree reconstruction can alter historical inferences concerning character evolution. For example, in the Pontederiaceae study, root position is uncertain and some near-optimal root positions alter the inferred polarity of floral characters. A second issue concerns the choice of character optimization scheme. The conclusion that heteromorphic incompatibility arose after floral trimorphism depends on an unequally, rather than an equally, weighted optimization scheme. Kohn et $al.$ (1996) defend their choice of unequal weighting ($\geqslant 2:1$) based primarily on the extreme rarity of trimorphic incompatibility in the angiosperms and its developmental and genetic complexity, which make its origin improbable relative to its evolutionary dissolution (see Barrett 1993). In contrast, with equal weighting Kohn et $al.$ (1996) found the highly unlikely pattern that tristyly evolved up to four times in Pontederiaceae. Virtually all phylogenetic studies on the evolution of plant reproductive traits have treated character shifts as equally likely, presumably because this procedure is simpler and avoids subjectivity involved in character weighting. However, with complex characters (e.g. floral syndromes such as tristyly) this may not be wise. When biological evidence indicates an unequal probability of possible transitions between reproductive traits, workers should at least explore how the character optimization employed influences historical reconstructions.

6 Concluding remarks

Comparative biology and phylogenetic reconstruction provide means for identifying and testing functional hypotheses; however, their reliance on phylogenetic trees currently limits their applicability to many issues in reproductive biology. Phylogenetic trees based on morphological and molecular characters both present problems for analysis of reproductive characters. Phylogenetic trees derived from morphology often rely heavily on floral traits, so that circularity can be a serious problem in studying the evolution of reproductive characters (Eckenwalder & Barrett 1986; Wyatt 1988). In contrast, phylogenies based on molecular characters often lack resolution at the species level. Hodges & Arnold (1994) reported a striking example of this problem in their study of adaptive radiation of pollination syndromes in *Aquilegia* (Ranunculaceae). Despite remarkable diversification of floral traits associated with the pollination systems of closely related species, little molecular divergence among the 27 species was evident in sequence variation from both nuclear (ITS) and chloroplast (*atb*B–*rbc*L spacer) regions. A major priority for biologists interested in phylogenetic reconstruction will be the search for genes that allow finer resolution at the species level. Until such genes are found, the availability of well resolved phylogenies will hamper comparative analysis of reproductive characters in some groups.

We thank Leigh Johnston and Mark Porter for providing phylogenetic data and information on the Polemoniaceae, Mike Dodd and Jonathan Silvertown for the MacClade version of the 'Chase phylogeny' of seed plants, Chris Eckert, Bill Cole, Kevin Fitzsimmons and Stephen Wright for help with data collection and analysis and Sean Graham, Richard Lenski, Dan Schoen, Jonathan Silvertown and Mark Westoby for helpful comments on the manuscript. Research funded by grants from the Natural Sciences and Engineering Reseach Council of Canada.

References

Barrett, S. C. H. (ed.) (1992). *Evolution and function of heterostyly*. Berlin: Springer-Verlag.
Barrett, S. C. H. (1993). The evolutionary biology of tristyly. *Oxford Surveys in Evolutionary Biology* **9**, 283–326.
Barrett, S. C. H. & Cruzan, M. B. (1994). Incompatibility in heterostylous plants. In *Genetic control of self-incompatibility and reproductive development in flowering plants*. (ed. E. G. Williams, A. E. Clarke & R. B. Knox). pp. 189–219. Dordrecht: Kluwer Academic Publishers.
Barrett, S. C. H. & Eckert, C. G. (1990). Variation and evolution of mating systems in seed plants. In *Biological approaches and evolutionary trends in plants*, (ed. S. Kawano). pp. 229–254. London: Academic Press.
Barrett, S. C. H. & Glover, D. E. (1985). On the Darwinian hypothesis of the adaptive significance of tristyly. *Evolution* **39**, 766–774.
Brooks, D. R., & McLennan, D. A. (1991). *Phylogeny, ecology and behavior: a research program in comparative biology*. University of Chicago Press.
Charlesworth, D. (1979). The evolution and breakdown of tristyly. *Evolution* **33**, 486–498.
Charlesworth, D. & Charlesworth, B. (1979). A model for the evolution of distyly. *American Naturalist* **114**, 467–498.

Charnov, E. 1982 *The theory of sex allocation*. Princeton University Press.

Chase, M. W. & 41 others (1993). Phylogenetics of seed plants: an analysis of nucleotide sequences from the plastid gene *rbc*L. *Annals of the Missouri Botanical Garden* **80**, 528–580.

Cruden, R. W. (1977). Pollen-ovule ratios: a conservative indicator of breeding systems in flowering plants. *Evolution* **31**, 32–46.

Cruzan, M. B., & Barrett, S. C. H. (1996). Post-pollination mechanisms influencing mating patterns and fecundity: an example from *Eichhornia paniculata*. *American Naturalist* **147**, 576–598.

Donoghue, M. J. (1989). Phylogenies and the analysis of evolutionary sequences, with examples from seed plants. *Evolution* **43**, 1137–1156.

Dressler, R. L. (1981). *The orchids: natural history and classification*. Cambridge, MA: Harvard University Press.

Eckenwalder, J. E., & Barrett, S. C. H. (1986). Phylogenetic systematics of Pontederiaceae. *Systematic Botany* **11**, 373–391.

Fægri, K. & van der Pijl, L. (1979). *The principles of pollination ecology*. Oxford: Pergamon Press.

Felsenstein, J. (1985). Phylogenies and the comparative method. *American Naturalist* **125**, 1–15.

Givnish, T. J. (1980). Ecological constraints on the evolution of breeding systems: dioecy and dispersal in gymnosperms. *Evolution* **34**, 959–972.

Gould, S. J., & Vrba, E. S. (1982). Exaptation: a missing term in the science of form. *Paleobiology* **8**, 4–15.

Graham, S. W. & Barrett, S. C. H. (1995). Phylogenetic systematics of Pontederiales: implications for breeding-system evolution. In *Monocotyledons: systematics and evolution* vol. 2 (eds. P. J. Rudall, P. J. Cribb, D. F. Cutler & C. J. Humphries). pp. 415–441. Kew: Royal Botanic Gardens.

Grant, V. (1959). *The natural history of the phlox family*. The Hague: Nijhoff.

Grant, V. & Grant, K. A. (1965). *Flower pollination in the phlox family*. New York: Columbia University Press.

Harvey, P. H. & Pagel, M. (1991). *The comparative method in evolutionary biology*. Oxford University Press.

Harvey, P. H., Read, A. A. & Nee, S. (1995). Why ecologists need to be phylogenetically challenged. *Journal of Ecology* **83**, 535–536.

Hodges, S. A. & Arnold, M. L. (1994). Columbines: a geographically widespread species flock. *Proceedings of the National Academy of Sciences, U.S.A.* **91**, 5129–5132.

Johnson, L. A., Schultz, L. L., Soltis, D. E. & Soltis, P. S. (1996). Monophyly and generic relationships of Polemoniaceae based on *mat*K sequences. *American Journal of Botany* **83**, 1207–1224.

Kirk, W. D. J. (1992). Interspecific size and number variation in pollen grains and seeds. *Biological Journal of the Linnean Society* **49**, 239–248.

Kohn, J. R. & Barrett, S. C. H. (1992). Experimental studies on the functional significance of heterostyly. *Evolution* **46**, 43–55.

Kohn, J. R., Graham, S. W., Morton, B., Doyle, J. J. & Barrett, S. C. H. (1996).

Reconstruction of the evolution of reproductive characters in Pontederiaceae using phylogenetic evidence from chloroplast DNA restriction-site variation. *Evolution.* **50**, 1454–1469.

Lloyd, D. G. (1980). Demographic factors and mating patterns in angiosperms. In *Demography and evolution in plant populations* (ed. O. T. Solbrig). pp. 67–88. Oxford: Blackwell.

Lloyd, D. G. & Webb, C. J. (1992). The selection of heterostyly. In *Evolution and function of heterostyly* (ed. S. C. H. Barrett), pp. 179–205. Berlin: Springer-Verlag.

Lord, E. M. & Eckhard, K. J. (1984). Incompatibility between the dimorphic flowers of *Collomia grandiflora*, a cleistogamous species. *Science* **223**, 695–696.

Maddison. W. P. & Maddison, D. R. (1992). *MacClade: analysis of phylogeny and character evolution.* Sunderland: Sinauer.

Plitmann, U. (1994). Assessing functional traits from herbarial material: the test case of pollen tubes in pistils of Polemoniaceae. *Taxon* **43**, 63–69.

Plitmann, U. & Levin, D. A. (1990). Breeding systems in the Polemoniaceae. *Plant Systematics and Evolution* **170**, 205–214.

Primack, R. B. (1979). Reproductive effort in annual and perennial species of *Plantago* (Plantaginaceae). *American Naturalist* **114**, 51–62.

Purvis, A. & Rambaut, A. (1995). Comparative analysis by independent contrasts (CAIC): an Apple Macintosh application for analysing comparative data. *Computer Applications in the Biosciences* **11**, 247–251.

Reeve, H. K. & Sherman, P. W. (1993). Adaptation and the goals of evolutionary research. *Quarterly Review of Biology* **68**, 1–32.

Stebbins, G. L. (1957). Self-fertilization and population variability in the higher plants. *American Naturalist* **91**, 337–354.

Stebbins, G. L. (1974). *Flowering plants: evolution above the species level.* Cambridge: Belknap Press.

Stuchlik, L. (1967). Pollen morphology in the Polemoniaceae. *Grana Palynol.* **7**, 146–240.

Swofford, D. L. (1991). *PAUP: phylogenetic analysis using parsimony, version* 3.1.1. Champaign, IL: Illinois Natural History Survey.

Uyenoyama, M. K., Holsinger, K. E. & Waller, D. M. (1993). Ecological and genetic factors directing the evolution of self-fertilization. *Oxford Surveys in Evolutionary Biology* **9**, 327–381.

Westoby, M., Leishman, M. R. & Lord, J. M. (1995). On misinterpreting the 'phylogenetic correction'. *Journal of Ecology* **83**, 531–534.

Williams, E. G., & Rouse, J. L. (1990). Relationships of pollen size, pistil length and pollen tube growth rates in *Rhododendron* and their influence on hybridization. *Sexual Plant Reproduction* **3**, 7–17.

Wright, S. (1977). *Evolution and the genetics of populations.* vol. 3 (*Experimental results and evolutionary deductions*). University of Chicago Press.

Wyatt, R. (1988). Phylogenetic aspects of the evolution of self-pollination. In *Plant evolutionary biology* (eds. L. D. Gottlieb & S. K. Jain). pp. 109–131. New York: Chapman & Hall.

5 · How does self-pollination evolve? Inferences from floral ecology and molecular genetic variation

Daniel J. Schoen, Martin T. Morgan and Thomas Bataillon

1 Introduction

The majority of angiosperms bear perfect flowers, i.e. flowers containing both anthers and stigmas. Most are outcrossing, but a significant proportion are predominantly self-pollinating. Indeed, the adoption of self-pollination is one of the most common trends in the evolutionary history of the angiosperms (Stebbins 1974).

Jain (1976) summarized hypotheses for the evolution of self-pollination. Two of most general are the 'automatic selection' and 'reproductive assurance' hypotheses. The automatic selection advantage of selfing arises because a gene promoting selfing in a population of outcrossers is, on average, transmitted to the next generation in two doses through the progeny arising from self-fertilization as well as an additional third dose through the male gametes that cross-fertilize ovules in the population, whereas the alternative gene for outcrossing is transmitted in only two doses (Fisher 1941). This transmission bias gives a strong advantage to mutations that increase the rate of selfing, which is negated only by inbreeding depression or by other correlates of selfing such as reduced male fertility (Holsinger *et al.* 1984). The reproductive assurance hypothesis, on the other hand, states that the selective advantage of self-pollination lies in assured seed production when pollinators are insufficient for full pollination of the ovules, e.g. due to poor climatic conditions or following long-distance dispersal to areas where pollinators or mates are absent (Baker 1955). Darwin (1876) believed that reproductive assurance is the chief reason for the evolution of selfing.

Our principal focus is on how theoretical and experimental approaches can be applied to determine the relative importance of automatic selection and reproductive assurance. We do this through examination of: (1) phenotypic selection models for the evolution of floral traits promoting selfing; and (2) theory of selectively neutral genetic variation, concentrating specifically on how patterns of neutral diversity are influenced by historical features accompanying the evolution of selfing.

2 Phenotypic selection of floral traits promoting self-pollination

2.1 The selection of selfing

The advantages of selfing proposed by the automatic selection and reproductive assurance hypotheses occur under opposing ecological conditions. Automatic selection is dependent on vector-mediated pollen transfer (for realization of the transmission bias), whereas reproductive assurance is manifested when conditions for vector-mediated pollen transfer are inadequate for full seed set. Under automatic selection, increased selfing evolves whenever the relative fitness of progeny from selfing is one half or greater than that of progeny from outcrossing (Fisher 1941). In contrast, autonomous selfing can evolve (through reproductive assurance) whenever lack of pollinators limits seed set, provided that progeny from selfing have at least some fitness and that selfing does not pre-empt ovules that would otherwise be cross-pollinated (Lloyd 1979). The automatic selection advantage has dominated much of the theoretical and experimental work on the evolution of self-pollination; e.g. as illustrated by the many studies of inbreeding depression in selfers and outcrossers. Moreover, the majority of models assume that selfing does not contribute to an increase in seed set (Fisher 1941; Nagylaki 1976; Wells 1979; Holsinger 1991), thereby minimizing any role for reproductive assurance. The models developed below are meant to examine the ecological conditions under which selfing is expected to evolve, with particular emphasis on how these conditions may influence opportunities for realization of the fitness gain.

2.2 Simultaneous changes in several modes of selfing

Lloyd (1979, 1992) demonstrated that how and when self-pollination occur during the lifetime of the flower (the 'mode' of selfing) strongly influences the conditions for its selection. For instance, when self-pollination depends upon the activities of pollinators ('facilitated self-pollination'), conditions for its selection differ from when it occurs independently of pollinators ('autonomous self-pollination'). This is true as well when selfing precedes or comes after opportunities for outcrossing (Lloyd 1979, 1992).

Our approach departs from that of Lloyd and others, who model the evolution of self-pollination *per se*, and thus consider only one mode of self-pollination at a time. Instead, we ask how floral changes that simultaneously influence the degree of *several* modes of selfing are selected. We assume that mutations causing changes in the timing of maturation or relative positions of anthers and stigmas within the flower are unlikely to influence only a single mode of selfing. Thus, in a self-compatible plant, a mutation that reduces the degree of dichogamy (temporal separation of anther and stigma maturation) or herkogamy (spatial separation of anther and stigma maturation) may lead to increased facilitated selfing. Such a mutation, however, may also lead to increased autonomous selfing. Likewise, a mutation that alters maturation times of the anthers and stigmas may lead to increased within-flower selfing, but such mutations may also increase the amount of between-flower (geitonogamous) selfing.

2.3 Selection of floral traits influencing autonomous and facilitated self-pollination

Assume a population is composed of two plant phenotypes that differ in the expression of floral trait z. The trait value z_i influences the amounts of both facilitated and autonomous self-pollination, $c(z_i)$ and $d(z_i)$, respectively, where the subscript i refers to the phenotype in question. The notation reminds us that the amount of facilitated and autonomous selfing are functions of the trait value z. Assume that autonomous selfing occurs after opportunities for outcrossing, i.e. autonomous selfing occurs via the 'delayed' autonomous mode of selfing described by Lloyd (1979), though our approach can be extended to other autonomous selfing modes. The proportion of ovules fertilized with the aid of an external vector (either through outcrossing or facilitated selfing) is symbolized by e, and the relative reduction in fitness of progeny from selfing versus outcrossing (inbreeding depression) is represented as δ. For simplicity, consider a population composed of a common phenotype 1, with floral trait expression z_1, and a rare phenotype 2 with floral trait expression, z_2 (phenotype 2 has the higher selfing rate). Facilitated selfing and outcrossing are each dependent on pollinators, and hence, the number of seeds produced through these processes is influenced by e. Autonomous (delayed) selfing, on the other hand, involves only that proportion of ovules remaining unfertilized after pollinator visitation, i.e. $1 - e$. With inbreeding depression, progeny from all modes of selfing have fitness $1 - \delta$ relative to progeny from outcrossing. Taking these factors together, we obtain the following expressions for female reproductive fitnesses:

$$W_{f1} = c(z_1)e(1 - \delta) + d(z_1)(1 - e)(1 - \delta) + [1 - c(z_1)]e$$
$$W_{f2} = c(z_2)e(1 - \delta) + d(z_2)(1 - e)(1 - \delta) + [1 - c(z_2)]e \qquad (1)$$

Male reproductive fitness is gained both through outcrossing and selfing. For the outcrossing component of male fitness, assume that phenotype 2 is rare, and hence, pollen from either phenotype encounters only ovules from phenotype 1. Male fitnesses gained through non-autonomous and autonomous self-pollination, and outcrossing may be written as:

$$W_{m1} = c(z_1)e(1 - \delta) + d(z_1)(1 - e)(1 - \delta) + [1 - c(z_1)]e$$
$$W_{m2} = c(z_2)e(1 - \delta) + d(z_2)(1 - e)(1 - \delta) + [1 - c(z_1)]e \qquad (2)$$

The relative fitness of phenotype 2 is:

$$\tilde{W}_2 = 1/2 \left[W_{f2}/\bar{W}_f + W_{m2}/\bar{W}_m \right] \qquad (3)$$

where $\bar{W}_f$ and $\bar{W}_m$ denote average fitness through female and male reproductive functions. To find the evolutionary stable floral trait value, z^*, solve for the conditions under which $dW_2/dz_2 = 0$ (when $z_1 = z_2 = z^*$), yielding the relationship:

$$2[c(z_2)'e(1 - \delta) + d(z_2)'(1 - e)(1 - \delta)] - c(z_2)'e = 0 \qquad (4)$$

where the apostrophes denote the derivatives of the functions $c(z_i)$ and $d(z_i)$.

Assume a linear relationship between z_i and the selfing rate under each mode. Note that the change in selfing caused by modifications to the floral trait may differ for each mode, so accordingly, $c(z_i) = m_c z_i + y_c$ and $d(z_i) = m_d z_i + y_d$, where m_c and m_d denote the respective slopes of the linear relationship between facilitated and autonomous selfing modes and the trait value, and y_c and y_d denote the respective y-intercepts. The mutant phenotype will invade the population whenever:

$$\delta < 1 - m_c e/\{2[m_c e + m_d(1 - e)]\} \qquad (5)$$

Figure 5.1 illustrates how pollinator activity and inbreeding depression levels specified in equation (5) combine to select for the floral trait. Conditions for the evolution of the trait are intermediate between those for strict facilitated selfing ($\delta < 1/2$) and those for autonomous (delayed) selfing ($\delta < 1$). The more important result, however, is that in pollinator-limited environments, when a change in floral trait expression leads to increases in *both* facilitated and autonomous selfing, even if the increase in autonomous selfing is small (i.e. $m_d/m_c < 1$), the selection of selfing is subject to more relaxed conditions compared with facilitated selfing alone (i.e. inbreeding depression is less of an

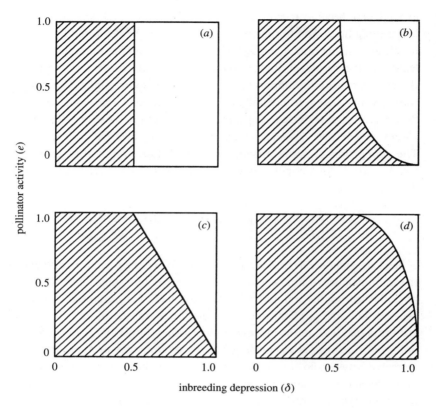

Figure 5.1. Conditions for the selection of a floral trait that causes a linear increase in both non-autonomous (slope m_c) and autonomous (slope m_d) self-pollination. Four cases: (a) $m_d/m_c = 0$; (b) $m_d/m_c = 0.1$; (c) $m_d/m_c = 1.0$; (d) $m_d/m_c = 3.0$. Hatched areas indicate parameter combinations where the trait is selected.

obstacle to the evolution of selfing). Reproductive assurance may, therefore, play an important role, not only in the evolution of autonomous selfing, but facilitated selfing as well.

2.4 Selection of floral traits influencing within- and between-flower self-pollination

Self-pollination may occur within as well as between flowers on the same plant, especially when it is facilitated by pollinators. Changes in flowers may influence the rates of both of these modes of selfing. For example, in a normally protandrous species, a mutation that causes the stigmatic lobes to expand and become receptive earlier on in floral development, thereby lengthening the period of time when self pollen can be deposited by insects, may lead to higher rates of selfing within the flower. But this same change also lengthens the period of stigma receptivity on other flowers of the plant, thereby increasing the opportunity for between-flower selfing. Consider a trait z, with value z_i in phenotype i, that influences within- and between-flower selfing, symbolized $c(z_i)$ and $g(z_i)$, respectively. Assume that both within- and between-flower selfing are facilitated by external pollen vectors. The proportion of seeds arising from self-fertilization is thus given by $[c(z_1) + g(z_1)]e$. In a population with phenotypes 1 and 2, the female reproductive fitnesses are:

$$W_{f1} = \{[c(z_1) + g(z_1)](1 - \delta) + [1 - c(z_1) - g(z_1)]\}e$$
$$W_{f2} = \{[c(z_2) + g(z_2)](1 - \delta) + [1 - c(z_2) - g(z_2)]\}e \qquad (6)$$

Next, consider male reproductive fitnesses. When the floral trait influences the amount of between-flower selfing, it is incorrect to assume that it will have no influence on the amount of pollen available for outcrossing. This is because the same mechanism of between-flower transfer of (self) pollen by vectors is that used in the process of outcrossing (Lloyd 1992). In the terminology of Holsinger et al. (1984), pollen discounting may be associated with increased between-flower selfing. Let $p(z_2)$ denote the proportion of a plant's pollen available for fertilizing outcrossed ovules in the population. Assume that phenotype 2 is rare, so that outcrossed pollen from either phenotype encounters only ovules of phenotype 1. This leads to the following male fitness expressions:

$$W_{m1} = \{[c(z_1) + g(z_1)](1 - \delta) + [1 - c(z_1) - g(z_1)]\}e$$
$$W_{m2} = \{[c(z_2) + g(z_2)](1 - \delta) + [1 - c(z_1)$$
$$- g(z_1)][p(z_2)/p(z_1)]\}e \qquad (7)$$

If one assumes that the tradeoff between the rate of between-flower selfing and the amount of pollen available for outcrossing is complete (Lloyd 1992), then $p(z_i) = 1 - g(z_i)$. Assuming, as before, that the rate of self-pollination is a linear function of the trait value z_i, we have $c(z_i) = m_c z_i + y_c$ and $g(z_i) = m_g z_i + y_g$. The evolutionary stable floral trait value satisfies $d\bar{W}_2/dz_2 = 0$ (when $z_1 = z_2 = z^*$), and it can be shown that the mutant phenotype will invade the population when:

$$\delta < 0.5 - km_g/[2(m_c + m_g)] \qquad (8)$$

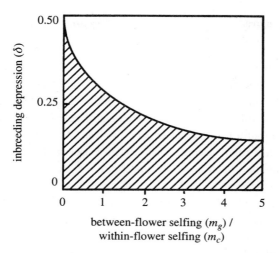

Figure 5.2. Conditions for the selection of a floral trait that causes a linear increase in both between- and within-flower self-pollination rates (slopes m_g and m_c, respectively) ($k = 0.75$). Hatched areas indicate parameter combinations where the trait is selected.

where k ($k \leqslant 1$) is a constant that reflects the proportion of total selfing in the wild type that is within-flower as opposed to between-flower. Values of k near 1 indicate that within-flower selfing predominates. Figure 5.2 illustrates the conditions for selection of a floral trait that influences both within- and between-flower selfing rates. When $m_g/m_c = 0$ (i.e. no effect of the trait on between-flower selfing), conditions for selection of increased selfing are identical to those for facilitated selfing ($d < 1/2$). But as the influence of the trait on between-flower selfing increases ($m_g/m_c > 0$), conditions for the evolution of the trait become more restrictive. The threshold values of δ shown in Figure 5.2 would be larger (less of an obstacle to the evolution of selfing) if pollen discounting accompanying between-flower selfing less severe, but the qualitative result shown there would still hold.

The results above indicate that increases in between-flower selfing, occurring in conjunction with increases in facilitated selfing, oppose the automatic-selection advantage. Consequently, automatic selection as a driving force for the evolution of selfing may be less important than simpler (one mode) models suggest.

2.5 Data and approaches pertaining to modes of selfing and reproductive assurance

Results from the selection models draw attention to the importance of understanding how and under which conditions self-pollination occurs. If facilitated and autonomous selfing often occur together, and pollinator activity is insufficient for full seed set, then a significant role for reproductive assurance in the evolution of selfing can be inferred. Also, if increases in facilitated selfing are often accompanied by increases in between-flower selfing, then automatic selection would appear to be less important than suggested by classical theory.

To gain increased understanding of the relative importance of automatic selection and reproductive assurance in the evolution of self-pollination, it would be helpful to obtain more information on the modes (or combination of modes) by which selfing occurs, as well as the amount of reproductive assurance it provides. The relative contributions of different modes of self-pollination can be studied by jointly manipulating flowers and estimating mating system parameters. For example, if the proportion of selfed ovules per plant, s, is due to a combination of within- and between-flower facilitated selfing, $s = c + g$, then overall selfing can be partitioned into these modes by measuring rates of selfing in two classes of seeds – those collected from intact (control) flowers and those collected from emasculated flowers. In intact flowers, outcrossing and all modes of selfing can occur. In emasculated flowers, only outcrossing and between-flower selfing can occur. The difference in estimated selfing rates between these two classes of seed provides a measure of the contribution of between-flower selfing to the overall selfing rate. These methods have been described in greater detail elsewhere (Schoen & Lloyd 1992).

What about estimates of the amount of reproductive assurance provided by selfing? Experimental procedures are quite simple, requiring only a comparison of seed production in intact flowers (where all modes of selfing, including autonomous selfing, can occur) with that of flowers emasculated at the start of anthesis (where autogamous selfing is prevented) under normal field conditions. Any increased seed production in intact flowers can be attributed to the reproductive assurance provided by selfing (this procedure assumes, of course, that emasculation does not reduce attractiveness of flowers to pollinators) (Schoen & Lloyd 1992).

Unfortunately, not much is known about the contributions of the different modes of selfing to the overall selfing rate, or the amount of reproductive assurance that selfing provides. While selfing rates have been estimated for many plant species (summary in Barrett & Eckert 1990), there have been only a few efforts to partitition selfing into component modes. For example, Leclerc-Potvin & Ritland (1994) found that in a population of *Mimulus guttatus* a 1:2 mixture of autonomous (within-flower) and geitonogamous selfing occurred. Schoen & Lloyd (1992) found that almost all selfing in the open (chasmogamous) flowers of the annual plant *Impatiens pallida* is due to geitonogamy. Estimates of selfing rates as a function of inflorescence size in experimental populations of *Malva moschata*, *Hibiscus moscheutos*, and *Eichhornia paniculata* (Snow *et al.* 1996) also suggest that geitonogamy may occur frequently.

With regard to the amount of reproductive assurance provided by selfing, Cruden & Lyon (1989) measured fruit set in emasculated and control flowers under natural field conditions in two species, *Calylopus serrulatus* and *Scilla sibirica*, and found that selfing increased seed output between 30% and 50% in the former species, but not at all in the latter. Piper *et al.* (1986) found that homostylous individuals of *Primula vulgaris* set significantly more seed than pins or thrums in the same populations. More extensive, but less direct, evidence that selfing can provide reproductive assurance comes from experiments that measure seed or fruit set in self-compatible plants isolated from pollinators. Such studies were recently summarized by Lloyd & Schoen (1992), and it was found that approximately three-fifths of the 37 species for which data were available produce some seed autonomously, but without complementary studies that directly assess reproductive assurance, as in Cruden & Lyon's (1989) investigation, we cannot yet be sure of the extent to which reproductive assurance through self-pollination is realized in nature.

3 Neutral gene diversity and the evolution of self-pollination

When self-pollination evolves by automatic selection, the way in which selfing spreads to other populations may differ from when it evolves by reproductive assurance (Figure 5.3). Specifically, as autonomous selfing allows for uniparental reproduction, one might expect that there will be more opportunities for the founding of populations by one or a few individuals (Baker 1955). The expectation of increased founding events does not accompany the evolution of selfing by automatic selection, as seed set remains dependent on the presence of pollinators and mates. These differences in population history may have a pronounced effect on the distribution of neutral diversity among populations within selfing species, as discussed below.

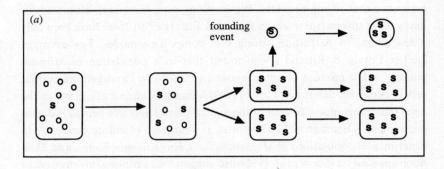

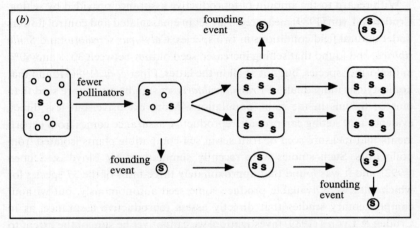

Figure 5.3. Hypothetical population histories associated with the selection of self-pollination. Under automatic selection (a) the mating system modifier spreads due to the transmission bias, but the probability of founding events is not enhanced. Under reproductive assurance (b), the mating system modifier spreads because it increases seed set in an environment where vector-mediated pollination is uncertain. Founding events are more likely due to uniparental reproduction. The symbols 's' and 'o' indicate selfing and outcrossing morphs.

3.1 Hitchhiking and the selection of mating system modifiers under automatic selection and reproductive assurance

The selection of a favourable mutation and its influence on variation at a linked neutral locus has been investigated by several researchers (Maynard Smith & Haigh 1974; Hedrick 1980). The issue of how the selection of a mating system modifier directly influences diversity at a neutral locus has not, however, been investigated. The question we pose here is whether the effect of hitchhiking differs under automatic selection versus reproductive assurance.

Consider a population with two diallelic loci A and B. Locus A determines the selfing rate, while locus B is selectively neutral. Let the frequencies of the ten two-locus genotypes be denoted as v_i, the frequencies of the gametes AB, Ab, aB, and ab as $x_{11}, x_{10}, x_{01}, x_{00}$, and the frequencies of alleles at loci A and B as p_1, q_1 and p_2, q_2. The selfing rates of genotypes AA, Aa and aa are s_1, s_2, and s_3, respectively, with $s_1 \leqslant s_2 < s_3$. For brevity, consider only the case where s_2 is exactly intermediate between s_1 and s_3 (codominance). Recursion equations for the v_i frequencies can be obtained by modifying those in Strobeck (1979), i.e. specifying the selfing rate for each genotype at the A locus.

Under automatic selection, the mating system modifier (a allele) will be introduced into the population at low frequency by mutation, and hence, linkage disequilibrium between the modifier and neutral locus will be near zero. Consider a population that is initially outcrossing. Figure 5.4a summarizes the effect of hitchhiking on diversity at neutral locus B as a function of the recombination fraction between the two loci when initial frequencies at the neutral locus B are $p_2 = q_2 = 0.5$ and the mating system modifier allele is introduced at $q_1 = 0.001$. Only when the effect of a on the selfing rate is large and the recombination fraction between the two loci is moderate to small (0.1 or less) is there any pronounced decrease in diversity at the neutral locus. If the influence of the a allele is reduced (e.g., from causing 100% selfing in the homozygote to only 10% selfing), the hitchhiking effect diminishes quite rapidly and is confined to a much smaller genomic region (Figure 5.4a). Hitchhiking has a more pronounced effect on neutral diversity when the mating system modifier is introduced into a population that is already partially selfing (Figure 5.4b), presumably because the selection of the modifier occurs more rapidly in the homozygous state.

To determine the effect of hitchhiking, assumptions must be made about the magnitude of selfing rate modification arising from floral change. Since most modifications to floral traits cause relatively small changes in the rate of selfing, it would appear from the analytical results that automatic selection will not lead to any broad (i.e. throughout the genome) reduction in initially outcrossing populations. A stronger (more genome-wide) effect on neutral diversity levels can, however, occur if such modifiers arise in partially selfing populations.

What about the effect of hitchhiking under reproductive assurance? As noted above, there is an increased likelihood of founding events in the history

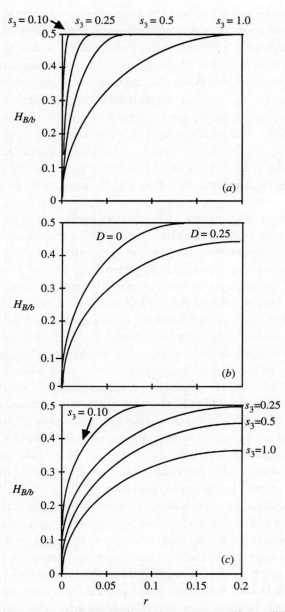

Figure 5.4. The effect of selection of a mating system modifier (a allele) on Nei's gene diversity at neutral, diallelic locus B ($H_{B/b}$) as a function of selfing rate (in the aa homozygote, s_3), and recombination fraction r. The mating system modifier is introduced into: (a) an outcrossing population in low frequency ($q_1 = 0.001$) and in

of a selfing species under this hypothesis, and thus an increased probability of linkage disequilibrium between the mating system modifier locus and any neutral loci remaining in a polymorphic state. This linkage disequilibrium is expected to enhance the decline in neutral diversity following a founding event (Hedrick 1980). For example, compare the situation in which the mating system modifier is introduced into an outcrossing population, in linkage equilibrium with the locus B (Figure 5.4a), to the case where a double heterozygote individual founds a new population – where linkage disequilibrium between the two loci is necessarily maximum (Figure 5.4c). The reduction in neutral diversity due to hitchhiking is more pronounced and extends to genomic regions more distant from the modifier (Figure 5.4c). The same effect can be seen when the ancestral (source) population is partially selfing (Figure 5.4b). The enhanced hitchhiking effect under reproductive assurance will contribute to increases in among-population variation in levels of neutral diversity (see below).

3.2 Indirect effects of selection of mating system modifiers on diversity at neutral loci

Will automatic selection and reproductive assurance have different indirect influences on diversity levels at neutral loci? First, note that inbreeding leads to a reduction in effective population size (N_e). The expected reduction is given by $N_e(selfer) = N_e(outcrosser)/(1 + F)$, where F is the inbreeding coefficient of the population arising from selfing (Pollak 1987). According to the sampling properties of the neutral model, (Ewens 1972), a reduction in effective population size due to inbreeding,will lead to a reduction in neutral allele diversity. Second, selfing leads to slower decay of linkage disequilibrium, and as noted in the case of hitchhiking between mating system loci and neutral alleles, this should contribute to reductions in neutral allelic diversity whenever any selected and neutral loci start out in linkage disequilibrium (Hedrick 1980). Third, with high levels of selfing, a reduction of neutral

near linkage equilibrium with locus B; (b) a population of partial selfers ($s_1 = 0.75$) in low frequency ($q_1 = 0.001$) and near zero linkage equilibrium with locus B ($D = 0$ curve), or into a founder population at high frequency ($q_1 = 0.5$) and in complete linkage disequilibrium with locus B ($D = 0.25$ curve) (the selfing rate of the aa homozygotes is $s_3 = 1.0$, both curves); and (c) a founder population at high frequency ($q_1 = 0.5$) and in complete linkage disequilibrium with locus B.

diversity may occur because of 'background selection', in which recurrent deleterious mutations located throughout the genome are selectively eliminated (Charlesworth *et al.* 1993). Fourth, when autonomous selfing increases the likelihood of bottlenecks, there is the expectation of reduced neutral diversity due to genetic drift (Nei *et al.* 1975; Tajima 1989).

Table 5.1 summarizes some expectations pertaining to patterns of neutral diversity under the two hypotheses for the evolution of selfing. Note that neutral diversity levels within single selfing populations are expected to be reduced by inbreeding effects on N_e and background selection, regardless of how selfing has evolved and, therefore, examination of average population levels of neutral diversity alone may not help to distinguish between automatic selection and reproductive assurance.

On the other hand, examination of variation in levels of neutral diversity *among* a number of populations of a selfing species has the potential to suggest the mechanism of evolution of selfing. In particular, species in which selfing has evolved by reproductive assurance are expected to consist of populations that have recently recovered from a bottleneck, together with older and more stable populations (Figure 5.3). Those populations that have recently passed through bottlenecks will have reduced diversity levels due to both drift and the enhancement of hitchhiking, whilst those populations that have been demographically stable for some time will store higher levels of neutral diversity. Thus, in species where selfing has evolved through reproductive assurance, one might expect significant among-population variation in neutral diversity. This expectation is not a necessary corollary of automatic selection.

3.3 Variation in allozyme diversity in the populations of selfing species

There have been many studies of allozyme variation in selfing plant species (Hamrick & Godt 1990; Schoen & Brown 1991) (Figure 5.5). Substantially more variation in allozyme diversity is found among populations of selfers compared with outcrossers. For example, most selfers surveyed consist of populations with no detectable allozyme diversity together with other populations having relatively high levels of diversity. Such a tendency towards L-shaped distributions of allozyme diversity within selfing species suggests a possible history of recent bottlenecks in some but not all populations, as envisioned under the reproductive assurance model. The inference is an indirect one, however, and is based on assumptions about neutrality of allozyme loci (Gillespie 1991). Moreover, L-shaped distributions of neutral

Table 5.1. *The effects of hitchhiking, background selection, and genetic drift on neutral gene diversity under two hypothetical mechanisms for the evolution of self-pollination*

Mechanism of evolution of self-pollination	Associated population history	Hitchhiking	Background selection	Genetic drift
Automatic selection	no association of the evolution of selfing with increased founding events	decline in diversity at neutral loci	decline in neutral diversity in all populations with high selfing rates	decline in diversity due to inbreeding effects on N_e
Reproductive assurance	the evolution of selfing coupled with population disturbance or founding events (bottlenecks)	as above, but additional bottlenecks increase the probability of linkage disequilibrium between selected and neutral loci, thereby contributing further to loss of neutral diversity	as above	as above, but additional bottlenecks contribute to loss of neutral diversity through genetic drift

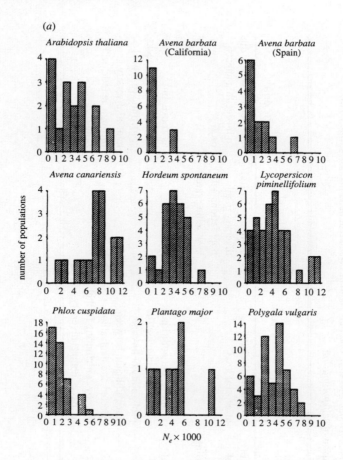

Figure 5.5. Variation among populations in allozyme diversity levels in (a) selfing and (b) outcrossing species, after Schoen & Brown (1991). Diversity is expressed as an estimate of effective population size (N_e) and is calculated from allelic richness at the sampled allozyme loci (Ewens, 1979).

variation could be produced if selfing populations within the species studied have evolved repeatedly from outcrossing ancestors, but at different times in the past, coupled with background selection (Charlesworth *et al.* 1993), i.e. selective sweeps that have removed neutral genetic variation in the more ancient selfing populations, but not in those in which selfing has recently evolved. Clearly, there is a need for methods that more directly reveal the effect of population history on neutral variation. One possible approach is examined next.

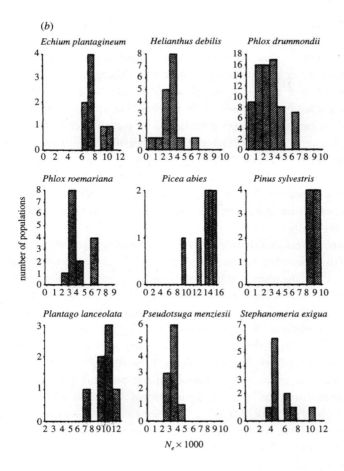

(b)

3.4 DNA sequence variation and the history of selfing populations

Here we focus on how coalescent theory might be applied to investigate historical events accompanying the evolutionary transition from outcrossing to self-fertilization. Of particular relevance is whether population bottlenecks occur in conjunction with this mating system transition, and how these bottlenecks influence the evolutionary history of a sample of genes. Slatkin & Hudson (1991) have used coalescent theory in an analogous manner to examine historical changes in population size.

Coalescent analyses characterize the statistical properties of a sample of genes by tracing the ancestors of these sampled genes backward over successive generations (Hudson 1990). A coalescent event occurs when two genes derive from a single ancestral gene in the previous generation. After many generations, the ancestry of all genes traces back to a single ancestral gene, the most recent common ancestor of the sample. The result of tracing the ancestry of sampled genes is a genealogy (Figure 5.6). Ecological factors such as population size and rate of self-fertilization determine the topology and length of branches in the genealogy. Mutations that accumulate along the branches of the genealogy provide information for statistical characterization of the sampled genes. Statistics include the number of segregating sites in the sample (Watterson 1975) and the average pairwise divergence between sampled genes (Tajima 1983). Since all sampled genes trace back to the most recent common ancestor, the time of the most recent common ancestor of the sampled genes sets an upper limit for the period about which evolutionary inferences can be drawn.

For instance, consider a sample of two genes in a diploid population of size N individuals with selfing rate s. For such a sample, there are two possible configurations for the genes – either both genes may be in the same individual, or each gene may be in separate individuals. Ecological parameters (e.g. population size, selfing rate) determine the probabilities of transition between different configurations (Milligan 1996). Consider two genes that in the present generation are in the same individual. Suppose that such an individual is derived from self fertilization, which occurs with probability s. In this case, the ancestors of the sampled genes in the previous generation (the 'parental' genes) must be in the same individual. The laws of Mendelian segregation tell us that half of the time both sampled genes are derived from the same parental gene (i.e. the sampled genes coalesce), and half of the time the sampled genes are derived from different genes which are nonetheless in the same individual.

On the other hand, suppose that the sampled genes are in an individual that arose from outcrossing, occurring with probability $1 - s$. The parental genes are in separate individuals with probability approximately equal to $1 - 1/N$, and in the same individual (because the same parental individual was randomly sampled twice) with probability $1/N$. If the genes are in the same individual, Mendelian rules of segregation tell us that there is probability $1/2$ that a coalescent event occurs, and probability $1/2$ that the genes are distinct but in the same individual. Combining the probabilities of self-fertilization and outcrossing, two genes in the same individual coalesce with probability $s(1/2) + (1 - s)(1/N)(1/2)$, are in the same individual *but do not coalesce* with probability $s(1/2) + (1 - s)(1/N)/(1/2)$, or are in different individuals with probability $(1 - s)(1 - 1/N)$. Changes in population size or

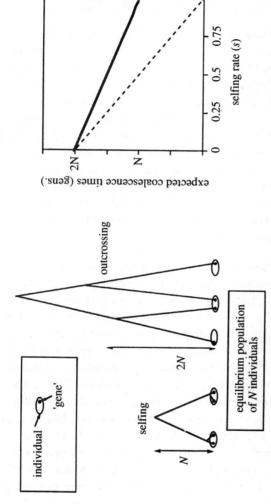

Figure 5.6. Hypothetical genealogies involving four sampled genes. In a completely selfing population, genes sampled within the same individual tend to coalesce early, whereas in an outcrossing population, genes sampled initially within the same individual coalesce (on average) at the same time as genes sampled in separate individuals. The accompanying graph shows average coalescence times for pairs of genes sampled within (broken line) and between (solid line) individuals as a function of the selfing rate in a population containing N individuals.

selfing rate alter these transition probabilities. Iteration of the transition probabilities allows straightforward determination of the genealogy of sampled genes, and hence the statistical properties of the sample.

3.5 The evolution of selfing and expected coalescent patterns

We describe results from an investigation of the historical change in rates of self-fertilization, reduction in population size, and the simultaneous effects of change in *both* selfing rate and population size. The models with population bottlenecks involve a reduction of populations to one tenth of their original size, persistence of the population at small size for a number of generations, and then expansion of the population to its original size. The timescale of all scenarios is in units of $\tau = N$ generations, so that the results presented do not depend on the exact population size of the simulations (only the relative size). The symbol T represents the period during which the bottlenecked populations remain small. For each scenario of ecological change, we calculated the expected time, t, to coalescence of two genes sampled in different individuals.

Several conclusions can be seen from the portion of this analysis that deals with historical changes in selfing.

1. The expected coalescent time of genes sampled within and between individuals decreases linearly with rate of self-fertilization (Figure 5.6). The decrease of expected coalescent times of genes sampled between individuals from $2N$ in a random mating population to N in a completely self-fertilizing population is comparable to the well-known linear increase in coefficient of consanguinity, just as the decrease in within-individual coalescent times parallels the increasing inbreeding coefficients found using traditional population genetic techniques (Crow & Kimura 1970).

2. The coalescent times of genes sampled within individuals in self-fertilizing populations provide little insight into the evolutionary history of populations because they coalesce quickly. On the other hand, genes sampled *between* individuals can potentially provide information on historical population parameters occurring over time spans of between $2N$ generations (in historically selfing populations) and $4N$ generations (in populations with historical outcrossing). This can be seen in Figure 5.7a, where the x-axis is the number of generations before the present, τ, when the population was last outcrossing and the y-axis is the expected between-individual coalescent time. The result illustrates the important point that demographic parameters such as the number of individuals in the popula-

tion impose an upper limit on the historical resolution attainable through sampling of genetic information in contemporary populations.

3. As the time of the transition from outcrossing to self- fertilization recedes into the past, the expected coalescent time changes monotonically from that characterizing outcrossing populations $(2N)$ to that characterizing completely self-fertilizing populations (N).

Next, consider the portion of the analysis that deals with historical reduction in population size (Figure 5.7b). It can be seen that historical bottlenecks occurring without any alteration in the rate of self-fertilization reduce expected between-individual coalescent times. Focussing on the solid line in Figure 5.7b (which corresponds to the coalescent in populations with intermediate self-fertilization and a bottleneck persisting for $T = 0.05N$ generations), we can see that when the bottleneck first occurred at a time τ less than the duration of the bottleneck, $\tau < T$, the contemporary population is at the reduced size and the coalescent times are those expected in a population suddenly reduced in size $T - \tau$ generations ago. Population size recovers from the bottleneck as τ becomes larger than T, and as the bottleneck recedes into history, the coalescent times gradually return to the value expected for the larger population. The main effect of population bottlenecks, then, is to reduce the between-individual coalescent times. The size of the reduction can be considerable and is proportional to the magnitude of the bottleneck, contrasting with the more modest effects of increase in self-fertilization that alter coalescent times by at most a factor of 2. Like the effects of changes in self-fertilization, though, demography imposes an upper limit on historical resolution of about $3N$ generations.

Finally, consider the evolution of self-fertilization occurring in conjunction with population bottlenecks. This leads to coalescent patterns that are approximately the additive combination of the separate effects of each individual factor (i.e. changes in self-fertilization, bottlenecks) (Figure 5.7c). When bottlenecks are not too severe (either of short duration or only small reduction in size), the consequences of past outcrossing express themselves in expected coalescent times greater than expected in the case of historically constant rates of selfing. Increasing the size or duration of the bottleneck overwhelms the effects of the evolution of self fertilization, leading to coalescent times much smaller than expected based on current population size and mating system.

These results hold some promise for inferring the mechanism of evolution of self fertilization based on observation of contemporary population struc-

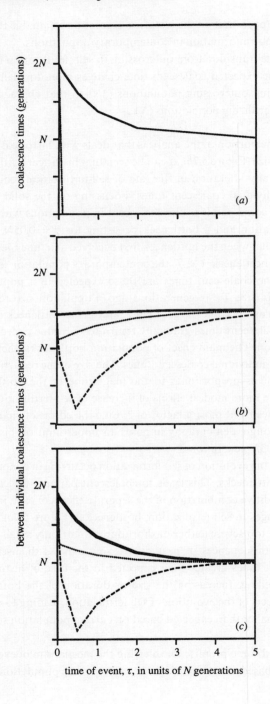

time of event, τ, in units of N generations

ture. If under the automatic selection hypothesis the transition from outcrossing to self-fertilization generally occurs in the absence of population bottleneck, then estimates of coalescent times and related statistics from contemporary populations may be characteristic of larger or more outcrossing populations than census and mating system estimates suggest. On the other hand, under reproductive assurance, coalescent times and derived statistics should reflect much smaller population sizes than census size would indicate.

There are important caveats. Population demographic parameters impose an upper time limit, of several N generations, during which mating system transitions are detectable. This restricts application to species where the mating system transition is thought to have occurred relatively recently. The effects of the selfing transition and of population bottleneck may cancel each other out so that, in the absence of additional information, the conclusion reached is simply that the population has experienced approximately constant selfing rates and population sizes for at least several N generations. Finally, all inferences drawn from contemporary samples rely on comparing observed with expected statistics (e.g. comparing coalescent times observed with those expected if the selfing rate and population size were constant). The inferential process presented above, therefore, relies on external information (e.g. additional information from sources other than the genetic data about contemporary population size and mating system). One way of circumventing this difficulty is through development of statistics that characterize the topology of the tree. For instance, Tajima (1989) shows that his D statistic, which contrasts the number of segregating sites with the average pairwise divergence, reflects recent changes in population size. Development of such statistics represent areas of active research.

Figure 5.7. Coalescence times for pair of genes sampled in a population: (a) that was historically outcrossing but where a transition to complete selfing occurred τN generations ago (N denotes population size and is assumed to be constant in this case; dotted line denotes within individuals; solid line denotes between individuals); (b) with constant selfing rate ($s = 0.5$), but where a bottleneck occurred τN generations ago reducing population size by a factor of 10 for a duration of TN generations (after which the population regained its original size N) (solid line denotes $T = 0.005N$; dotted line denotes $T = 0.05N$; broken line denotes $T = 0.5N$); (c) where a transition from historical outcrossing to complete selfing occurred τN generation ago in conjunction with a bottleneck reducing the population size by a factor of 10 for a duration of TN generations (after which the population regained its original size N) (solid line denotes $T = 0.005N$; dotted line denotes $T = 0.05N$; broken line denotes $T = 0.5N$).

We thank Spencer Barrett, Kent Holsinger, Russell Lande and Stewart Schultz for discussing the material presented here.

References

Baker, H. G. (1955). Self-compatibility and establishment after 'long-distance' dispersal. *Evolution* 9, 347–348.

Barrett, S. C. H. & Eckert, C. G. (1990). Variation and evolution of plant mating systems. In *Biological Approaches and Evolutionary Trends in Plants* (ed. S. Kawano). New York: Academic Press.

Charlesworth, B., Morgan, M. T. & Charlesworth, D. (1993). The effect of deleterious mutations on neutral molecular variation. *Genetics* **134**, 1289–1303.

Crow, J. F. & Kimura, M. (1970). *An introduction to population genetics theory.* New York: Harper and Row.

Darwin, C. (1876). *The effects of cross- and self-fertilization in the vegetable kingdom.* London: John Murray.

Ewens, W. J. (1972). The sampling theory for selectively neutral alleles. *Theoretical Population Biology* 3, 87–112.

Fisher, R. A. (1941). Average excess and average effect of a gene substitution. *Annals of Eugenics* **11**, 53–63.

Gillespie, J. H. (1991). *The causes of molecular evolution.* Oxford University Press.

Hamrick, J. L. & Godt, M.J. (1990). Allozyme diversity in plant species. In *Plant population genetics, breeding, and genetic resources.* (eds. A. H. D. Brown, M. T. Clegg, A. L. Kahler & B. S. Weir), pp. 43–63. Sunderland: Sinauer.

Hedrick, P. W. (1980). Hitch-hiking: a comparison of linkage and partial selfing. *Genetics* **94**, 791–808.

Holsinger, K. E. (1991). Mass-action models of plant mating systems: the evolutionary stability of mixed mating systems. *American Naturalist* **138**, 606–622.

Holsinger, K. E., Feldman, M. W. & Christiansen F. B. (1984). The evolution of self-fertilization in plants. *American Naturalist* **124**, 446–453.

Hudson, R. R. (1990). Gene genealogies and the coalescent process. *Oxford Surveys in Evolutionary Biology* 9, 1–44.

Jain, S. K. (1976). The evolution of inbreeding in plants. *Annual Review of Ecology and Systematics* 7, 469–495.

Lloyd, D. G. (1979). Some reproductive factors affecting the selection of self-fertilization in plants. *American Naturalist* **113**, 67–69.

Lloyd, D. G. (1992). Self- and cross-fertilization in plants. II. The selection of self-fertilization. *International Journal of Plant Science* **153**, 370–380.

Maynard Smith, J. & Haigh, J. (1974). The hitch-hiking effect of a favorable gene. *Genetical Research* **231**, 1114–1116.

Milligan, B. G. (1996). Estimating long-term mating systems using DNA sequences. *Genetics* **142**, 619–627.

Nagylaki, T. (1976). A model for the evolution of self-fertilization and vegetative reproduction. *Journal of Theoretical Biology* **58**, 55–58.

Nei, M., Maruyama, T. & Chakraborty, R. (1975). The bottleneck effect and genetic variability in populations. *Evolution* **29**, 1–10.

Pollak, E. (1987). On the theory of partially inbreeding finite populations. I. Partial selfing. *Genetics* **117**, 353–360.

Schoen, D. J. & Brown, A. H. D. (1991). Interspecific variation in population gene diversity and effective population size correlates with the mating system in plants. *Proceedings of the National Academy of Sciences of the U.S.A.* **88**, 4494–4497.

Slatkin, M. & Hudson, R. R. (1991). Pairwise comparison of mitochondrial DNA sequences in stable and exponentially growing populations. *Genetics* **129**, 555–562.

Stebbins, G. L. (1974). *Flower plants: evolution above the species level.* Cambridge, MA: Harvard University Press.

Strobeck, C. (1979). Partial selfing and linkage: the effect of a heterotic locus on a neutral locus. *Genetics* **92**, 305–315.

Tajima, F. (1983). Evolutionary relationship of DNA sequences in finite populations. *Genetics* **105**, 437–460.

Tajima, F. (1989). The effect of change in population size on DNA polymorphism. *Genetics* **123**, 597–601.

Watterson, G. (1975). On the number of segregating sites in genetic models without recombination. *Theoretical Population Biology* **7**, 256–276.

Wells, H. (1979). Self-fertilization: advantageous or disadvantageous? *Evolution* **33**, 252–255.

6 · Effects of life history traits on genetic diversity in plant species

J.L. Hamrick and M.J.W. Godt

1 Introduction

The availability of biochemical and molecular techniques to identify single-gene markers has made it possible to estimate genetic diversity for a wide array of organisms. In particular, the use of electrophoretic techniques to estimate genetic diversity at allozyme loci has produced a rich source of comparative data on the genetic diversity contained within plant and animal species. Currently more than 2200 studies have reported allozyme variation for seed plants. Allozyme loci are particularly valuable for comparative population genetic studies as they are fairly numerous and are codominant. Moreover, the same enzyme systems can usually be resolved in different plant taxa.

Within ten years of the first plant allozyme studies, plant population geneticists began to summarize the data available for seed plants to determine whether generalizations could be made concerning genetic diversity and its distribution. These early reviews revealed that plant species with different breeding systems, seed dispersal mechanisms, geographic ranges and life forms tended to maintain different mean levels of genetic diversity within their populations (Gottlieb 1977; Hamrick 1978; Brown 1979; Hamrick *et al.* 1979). Subsequent reviews compared levels of genetic diversity among populations within species and among related species. Interspecific comparisons demonstrated that genetic distance statistics were generally predictive of phylogenetic relationships. For example, progenitor-derived species pairs tended to be more genetically distinct than populations within species but less genetically distinct than well-defined congeners (Gottlieb 1977; Crawford 1983). Hamrick (1983) and Loveless & Hamrick (1984) used several life history and ecological traits to determine whether interpopulation genetic heterogeneity was related to the species' characteristics. They found that life form, geographic range, breeding system and taxonomic status had significant effects on the partitioning of genetic diversity within and among plant populations.

The most comprehensive review of plant allozyme diversity is that of Hamrick and Godt (1989). This review was the first to simultaneously calculate genetic diversity parameters within *and* among populations. Parameters were also introduced to describe genetic diversity within species (typically only within-population parameters had been reported). These analyses indicated that significant differences in genetic diversity existed between species with different life history traits. An additional observation was that the majority (about 75%) of the variation in genetic diversity among species was not explained by the eight life history traits. However, these traits explained nearly 50% of the interspecific variation in among-population genetic diversity.

One shortcoming of the previous reviews was that comparisons between species categorized for combinations of several life history traits were precluded by the limited number of entries in the database. For example, it was not possible to determine directly whether outcrossing species with large geographic ranges had more genetic diversity than selfing, endemic species.

Since our last review (prepared in 1988) the number of plant allozyme studies has increased by approximately 250 per year. The number of studies in our data base now permits analyses of two-trait combinations of the life history traits. In this chapter we examine genetic diversity in seven two-trait combinations that earlier analyses indicated are most likely to influence genetic diversity and its distribution. Among groups, we highlight traits that are associated with extremes of genetic variation and its distribution. By examining relationships between two life history traits and genetic diversity and its distribution, we hoped to identify significant biological patterns. The current size of our database also allowed us to compare genetic diversity parameters for several plant families.

2 Methods

The data utilized in this review is an updated (to autumn 1992) version of the data used by Hamrick & Godt (1989). Only those studies with genetic interpretations of electrophoretic banding patterns were included in the analyses. For each paper and species (i.e. an entry) we extracted or calculated three parameters used to measure genetic diversity within species and among the species' populations. To be included, studies were required to report data from polymorphic and monomorphic loci or report data of polymorphic loci from at least two populations. Deficiencies in the data for a relatively high proportion of the papers precluded the calculation of all genetic diversity parameters from many studies. Taxa that were the focus of more than one study (e.g. *Pinus sylvestris*) are represented more than once in the data. We chose this approach, rather than calculating a mean value for each species represented by multiple entries, because different entries often contain unique information. For example, different entries often represent population samples from different sections of the species' range and/or have utilized different loci. Due to the overall size of the database the redundancy caused by including multiple entries per species has little effect on the mean values for any group. There were 1491 entries considered for this review; on average 735 entries supplied useful data.

2.1 Genetic diversity parameters

Three genetic diversity parameters were calculated: the percentage of loci polymorphic within the species (P_s), genetic diversity within species (H_{es} = Hardy–Weinberg expected heterozygosity, Weir 1990), and the proportion of total genetic diversity residing among populations (G_{ST}). The proportion of polymorphic loci was calculated by dividing the number of loci polymorphic within the species (i.e. entry) as a whole by the total number of loci analysed. Genetic diversity was calculated for each locus (including monomorphic and polymorphic loci) by:

$$H_{es} = 1 - \sum \bar{p}_i^2$$

where $\bar{p}_i$ is the frequency of the *i*th allele pooled across all populations analysed for the species. Mean genetic diversity at the species level was obtained by averaging H_{es} over all loci. Variation among populations was estimated with Nei's genetic diversity statistics (Nei 1973). Total genetic diversity for the species (H_T) and mean diversity within populations ($\bar{H}_S$) were estimated for each polymorphic locus using the equation given above. The proportion of genetic diversity residing among populations (G_{ST}) was determined for each polymorphic locus by:

$$G_{ST} = (H_T - \bar{H}_S)/H_T$$

Mean H_T, $\bar{H}_S$ and G_{ST} values were calculated by averaging values obtained for each polymorphic locus.

We have chosen to report genetic diversity statistics calculated for the species rather than the more commonly used within-population values. Our rationale is that the measurement of genetic diversity within species is more biologically meaningful than mean population genetic diversity. Moreover, as species with the same overall genetic variation may have quite different mean within-population genetic diversities (depending on the distribution of genetic diversity among populations), we conclude that species values of P_s and H_{es} coupled with estimates of G_{ST} provide the most succinct and informative descriptors of genetic diversity.

2.2 Combinations of life history traits

In our earlier review (Hamrick and Godt 1989) we classified each species using categorical variables for each of eight life history traits: taxonomic status, life form, geographic range, regional distribution, breeding system, seed dispersal, mode of reproduction and successional status. The classification of each species for these eight traits was determined from descriptions in the original papers or from floras. On the basis of the results of our previous analyses, we limited the current analysis to the five traits that had the greatest influence on the levels and distribution of genetic diversity. These were: (1) breeding system, (2) seed dispersal mechanism, (3) life form, (4) geographic range and (5) taxonomic status (Table 6.1). All six combinations of traits 1–4 were analysed, plus breeding system × taxonomic status (see Table 6.2 for the complete list). When fewer than 15 entries provided useful data for a two-trait category, the category was excluded from the analyses. A slightly lower cut-off ($n > 13$) was used for the comparison of genetic diversity among plant families.

2.3 Statistical analyses

The statistical analyses employed generally followed procedures utilized by our earlier review (i.e. Hamrick & Godt 1989). Means and standard errors of genetic diversity parameters were calculated for each two-trait category. Differences among categories of the seven combinations were analysed by performing separate one-way ANOVAs using the GLM procedures of SAS (SAS Institute, Inc. 1988) with trait categories treated as class variables. A least-squares means procedure (SAS Institute, Inc. 1988) was employed in a pairwise fashion to determine significant differences among categories.

Table 6.1. *Categories of each of the five life history traits used to produce the two-trait combinations*

(See Hamrick & Godt 1989 for more complete explanation of the traits.)

Breeding system	Seed dispersal mechanism	Life form	Geographic range	Taxonomic status
outcrossing	attached	annual	endemic	gymnosperm
mixed mating	gravity	short-lived	narrow	dicotyledon
selfing	ingested	perennial	regional	monocotyledon
	wind	long-lived perennial	widespread	

Table 6.2. *Proportion of variation among species (R^2) explained by each combination of traits*

Trait combination	R^2		
	P_s(%)	H_{es}	G_{ST}
Breeding system/seed dispersal mechanism	0.159	0.110	0.392
Breeding system/taxonomic status	0.130	0.069	0.390
Breeding system/geographic range	0.112	0.087	0.370
Life form/breeding system	0.148	0.089	0.396
Life form/seed dispersal mechanism	0.145	0.087	0.222
Life form/geographic range	0.121	0.101	0.264
Seed dispersal mechanism/geographic range	0.120	0.107	0.126

3 Results and discussion

The ANOVAs performed on the three genetic diversity parameters revealed highly significant differences ($p < 0.0001$) among categories of the seven two-trait combinations. However, the proportion of interspecific variation (R^2) in P_s and H_{es} explained by the analysis was quite low (Table 6.2). The R^2 values for P_s ranged from 0.112 to 0.159 while R^2 values for H_{es} were lower (0.069–0.110). Thus, most of the variation in genetic diversity among species was not explained by the two-trait combinations. This was not surprising as only 24% of the interspecific variation in H_{es} was accounted for by a multiple regression model that incorporated eight life history traits (Hamrick & Godt 1989). In the 1989 analysis, geographic range accounted for 32% of the explained variation while life form, breeding system and seed dispersal also explained significant proportions (25%, 17% and 17%, respectively).

A much higher proportion of the overall interspecific variation for G_{ST} was explained by the two-trait categories (Table 6.2), with R^2 values for G_{ST} ranging from 0.126 to 0.396. Trait combinations that involved the breeding system explained a relatively high proportion of the variation (mean $R^2 = 39\%$). Trait combinations involving life form, seed dispersal and geographic range had moderate R^2 values (mean $R^2 = 28\%$, 25% and 25%, respectively). In our earlier analyses the eight life history traits accounted for 47% of the heterogeneity in G_{ST} values (Hamrick & Godt 1989). Breeding system and life form were most closely associated with the among-species variation in G_{ST} and together accounted for 84% of the variation explained (Hamrick & Godt 1989). We have argued previously (Hamrick & Godt 1989, 1996a) that the genetic diversity maintained by a species is a function not only of its life history traits but also depends heavily on the species' ecological and evolutionary history. Fluctuations in the number and size of populations, biogeography and the speciation process itself may have played critical roles in determining the current genetic composition of species.

3.1 Analyses of two-trait combinations

Breeding system and seed dispersal

The trait combinations involving breeding systems and seed dispersal mechanisms explained relatively high proportions of the interspecific variation in P_s, H_{es} and G_{ST} (Table 6.2). For the percentage of polymorphic loci (P_s), three combinations (outcrossing/attached, outcrossing/wind and selfing/attached) had significantly higher values (Table 6.3). Selfing, gravity-dispersed species had the lowest mean values. A similar result was seen for H_{es}; selfing species with animal-attached seeds had the highest genetic diversity and selfing, gravity-dispersed species had the lowest H_{es} values. Species with different life history traits had markedly different G_{ST} values. On average, selfing species with gravity-dispersed seeds exhibited five-fold more differentiation than outcrossing, wind-dispersed species.

When seed dispersal mechanism was held constant and breeding system categories were varied, we found only a weak pattern for P_s and H_{es}. Although comparisons were difficult due to missing categories, it appears that for gravity-dispersed species, P_s and H_{es} were significantly lower in selfing species. Breeding system categories had a much more pronounced effect on G_{ST}. Gravity-dispersed, outcrossing species had significantly lower G_{ST} values than either mixed-mating or selfing species with gravity-dispersed seeds.

Table 6.3. *Mean levels of genetic variation within species and its distribution among populations for combined categories of breeding system and seed dispersal mechanism*

(n = mean number of entries, P_s = percentage of loci polymorphic within species, H_{es} = genetic diversity within species, G_{ST} = proportion of total genetic diversity at polymorphic loci found among populations. See the text for an explanation of these parameters. Values that do not share a superscripted letter are significantly different at $p < 0.05$.)

Trait combination	n	$P_s(\%)$	H_{es}	G_{ST}
Outcrossing				
attached	63	67.9[a]	0.188[b]	0.114[d]
gravity	178	50.2[b]	0.152[cde]	0.189[c]
ingested	54	52.4[b]	0.200[ab]	0.223[c]
wind	186	62.4[a]	0.157[cd]	0.101[d]
Mixed-mating				
gravity	63	52.7[b]	0.174[bc]	0.248[c]
ingested	17	34.1[c]	0.108[ef]	0.269[c]
wind	62	42.0[bc]	0.118[ef]	0.175[cd]
Selfing				
attached	29	64.7[a]	0.236[a]	0.426[b]
gravity	94	34.5[c]	0.097[f]	0.533[a]

Species with attached- or wind-dispersed seeds show a similar pattern across breeding system classes.

When species with the same breeding system but different seed dispersal mechanisms were considered, we found that outcrossing species with attached or wind-dispersed seeds had lower mean G_{ST} values than outcrossing species with gravity or ingested seed dispersal mechanisms. Within mixed-mating species, wind-dispersed species also had a somewhat lower G_{ST} value, while species with gravity and ingested seed dispersal mechanisms had similar values. Among selfers, gravity-dispersed species had significantly higher G_{ST} values than species that disperse seeds via attachment. Thus, species that we perceive to have limited pollen and seed dispersal tend to have more genetic differentiation among populations than species with more potential for gene movement. No such patterns were seen for genetic diversity within species.

Breeding system and taxonomic status
Comparisons between breeding system and the taxonomic status of species explained relatively little of the interspecific variation for P_s and H_{es} but

Table 6.4. *Mean levels of genetic variation within species and its distribution among populations for combined categories of breeding system and taxonomic status*

(See Table 6.3 for definitions of the symbols used.)

Trait combination	n	$P_s(\%)$	H_{es}	G_{ST}
Outcrossing				
gymnosperm	115	70.8[a]	0.169[ab]	0.073[e]
monocot	78	52.5[b]	0.158[ab]	0.157[d]
dicot	286	54.0[b]	0.165[ab]	0.184[d]
Mixed-mating				
monocot	20	53.1[b]	0.183[ab]	0.212[cd]
dicot	94	46.6[b]	0.143[ab]	0.240[c]
Selfing				
monocot	40	55.4[b]	0.195[a]	0.412[b]
dicot	98	32.6[c]	0.091[c]	0.587[a]

explained a relatively high proportion of the variation for G_{ST} (Table 6.2). Selfing dicots had the lowest P_s and H_{es} values and the highest G_{ST} value, while outcrossing gymnosperms had the highest P_s value and a G_{ST} value that was eight-fold lower than that for selfing dicots (Table 6.4). Selfing monocots had the highest H_{es} value. The high H_{es} values seen for selfing monocots may be due to the relatively high number of crop species (24%) represented in this category. A recent review of the crop allozyme literature indicated that monocot crops have elevated levels of genetic diversity while dicot crops have values that are equivalent to non-crop dicots (Hamrick & Godt 1996*b*). Within outcrossing species gymnosperms had significantly lower G_{ST} values than the angiosperms. There were no significant differences for H_{es} among the outcrossing taxa. There was a weak trend in the data indicating that mixed-mating or selfing monocots may have more genetic diversity than dicots with the same breeding systems. Monocots had lower G_{ST} values than dicots. The high proportion of wind-pollinated species within the monocots (91 of 138 monocot taxa were grasses) may have influenced this result.

Within taxonomic categories the various breeding systems produced a predictable pattern. As expected, both outcrossing monocots and dicots had lower G_{ST} values than monocots or dicots that self or have a mixed-mating system. Outcrossing dicots generally had more genetic diversity than dicots with mixed-mating or selfing mating systems. This trend is reversed in the monocots with selfing monocots having somewhat higher H_{es} values than

Table 6.5. *Mean levels of genetic variation within species and its distribution among populations for combined categories of breeding system and geographic range*

(See Table 6.3 for definitions of the symbols used.)

Trait combination	n	$P_s(\%)$	H_{es}	G_{ST}
Outcrossing				
endemic	57	54.4[ab]	0.142[cdef]	0.179[de]
narrow	131	55.8[ab]	0.155[bcd]	0.169[de]
regional	211	59.9[a]	0.171[abc]	0.120[e]
widespread	79	55.3[ab]	0.183[ab]	0.170[de]
Mixed-mating				
endemic	24	40.5[cd]	0.100[ef]	0.174[de]
narrow	33	40.5[cd]	0.123[def]	0.326[e]
regional	30	49.3[abc]	0.164[abcd]	0.272[cd]
widespread	27	60.1[a]	0.206[a]	0.169[de]
Selfing				
endemic	16	13.5[e]	0.034[g]	0.591[a]
narrow	19	28.7[de]	0.093[fg]	0.512[ab]
regional	68	45.1[c]	0.121[def]	0.572[a]
widespread	38	47.2[bc]	0.165[abcd]	0.446[b]

either mixed-mating or outcrossing species. The different patterns seen for the monocots and dicots may be due to the fact that many outcrossing dicots are woody plants while outcrossing monocots are typically herbaceous. Hamrick & Godt (1989) and Hamrick *et al.* (1992) have shown that, regardless of their taxonomic status, woody plants have significantly higher P_s and H_{es} values than either annuals or perennial herbaceous species.

Breeding system and geographic range

The combination of breeding system and geographic range explained little of the variation in genetic diversity within species (H_{es}) but a relatively high proportion of the variation among species for G_{ST} (Table 6.2). The highest P_S values were found for outcrossing, regionally distributed species and mixed-mating widespread species (Table 6.5). Widespread, mixed-mating species also had the highest H_{es} value. The lowest H_{es} and P_s values were for endemic selfing species. The lowest G_{ST} value was found for outcrossing, regionally distributed species, while the highest G_{ST} value was for endemic, selfing species. Selfing endemic species had the lowest P_s and H_{es} and the highest G_{ST} values for any two-trait category in our overall analysis.

Within a breeding system category, endemic and narrowly distributed species tended to have lower P_s and H_{es} values than species with more extensive geographic ranges. There was, however, no such pattern for G_{ST}. Endemic and narrowly distributed species have G_{ST} values that were equivalent or smaller than those of more widely distributed species.

Within each geographic distribution category outcrossing species tended to have lower G_{ST} values than either mixed-mating or selfing species. However, the difference in G_{ST} seems to be greatest between mixed-mating and selfing species. For endemic and narrowly distributed species there was a gradual decrease in H_{es} values with increased inbreeding, but no distinct trend was seen for widespread species. By our definition, widespread species occur on two or more continents. This group is generally biased towards weedy species whose human-oriented ecology may influence genetic diversity and its distribution.

Life form and breeding system
The combination of life form and breeding system also explained low proportions of the variation found among species for P_s and H_{es} but had the highest R^2 value for G_{ST} (Table 6.2). Long-lived perennials with an outcrossing breeding system had the highest P_s value, while annual outcrossing species had the highest H_{es} values (Table 6.6). Short-lived perennial species with selfing breeding systems had the lowest P_s and H_{es} values. The range in G_{ST} values was portrayed by the five-fold difference in G_{ST} between annual selfing species and long-lived, outcrossing species.

Within the life form categories there was a decrease in genetic diversity and an increase in G_{ST} with increased selfing. Within the breeding system categories there was a significant decrease in G_{ST} values with increasing longevity. For each of the three breeding system categories, G_{ST} was highest for annual plants, intermediate for short-lived perennials and lowest for long-lived perennials. No general pattern was apparent for P_s and H_{es}. The effect of life form on G_{ST} was probably due to differences in the pollen and seed dispersal abilities between herbaceous and woody plants. The larger stature and relatively lower population densities characteristic of trees should result in more gene dispersal (and, thus, lower G_{ST} values).

Life form and seed dispersal
Combinations of life forms and seed dispersal mechanisms explained relatively low proportions of interspecific variation in P_s and H_{es} and a lower proportion of variation in G_{ST} than did combinations that included the

Table 6.6. *Mean levels of genetic variation within species and its distribution among populations for combined categories of life form and breeding system*

(See Table 6.3 for definitions of the symbols used.)

Trait combination	n	$P_s(\%)$	H_{es}	G_{ST}
Annual				
outcrossing	98	59.1[b]	0.186[a]	0.191[cd]
mixed-mating	29	40.3[ef]	0.115[cd]	0.343[b]
selfing	102	43.2[de]	0.131[c]	0.553[a]
Short-lived perennial				
outcrossing	140	43.7[cd]	0.123[c]	0.218[cd]
mixed-mating	48	53.6[bc]	0.172[ab]	0.238[c]
selfing	32	29.9[f]	0.081[d]	0.442[b]
Long-lived perennial				
outcrossing	241	65.5[a]	0.180[ab]	0.094[e]
mixed-mating	24	42.5[ef]	0.135[bc]	0.145[de]

Table 6.7. *Mean levels of genetic variation within species and its distribution among populations for combined categories of life form and seed dispersal mechanism*

(See Table 6.3 for definitions of the symbols used.)

Trait combination	n	$P_s(\%)$	H_{es}	G_{ST}
Annual				
attached	42	74.2[a]	0.227[a]	0.277[bc]
gravity	141	44.9[c]	0.132[de]	0.380[ab]
ingested	26	34.8[d]	0.138[cde]	0.406[a]
wind	21	51.8[bc]	0.156[cbd]	0.392[ab]
Short-lived perennial				
attached	26	55.7[b]	0.165[cbd]	0.230[cd]
gravity	144	43.8[cd]	0.133[de]	0.233[cd]
ingested	21	38.3[cd]	0.128[cde]	0.321[abc]
wind	43	40.7[cd]	0.100[e]	0.266[cd]
Long-lived perennial				
attached	33	63.8[ab]	0.185[ab]	0.094[ef]
gravity	50	58.0[b]	0.178[c]	0.177[de]
ingested	28	64.8[ab]	0.225[a]	0.099[def]
wind	159	63.8[b]	0.159[cbd]	0.086[f]

breeding system (Table 6.2). Annual species with attached dispersal mechanisms have the highest values of P_s and H_{es}, while annual plants with ingested seeds have the lowest P_s value, and short-lived perennial species with wind-dispersed seeds have the lowest H_{es} value (Table 6.7). The highest G_{ST} values were found for annual plants with ingested seeds, while the lowest G_{ST} values were produced by long-lived perennials with wind-dispersed seeds.

Within the life form categories, species that disperse seeds by animal attachment had higher P_s and H_{es} values than species with other seed dispersal mechanisms, although the differences were not large. For G_{ST} there was no discernible pattern between seed dispersal mechanisms. Within seed dispersal categories there was no recognizable pattern for P_s across life forms, but there was a weak pattern for H_{es}. Longer-lived plants tended to have higher H_{es} values than either annual or short-lived species regardless of their seed dispersal mechanisms. This trend was particularly marked for species with gravity and ingested seed dispersal mechanisms. In contrast, G_{ST} values were strongly influenced by life form. For all seed dispersal categories, annual plants had approximately three-fold higher G_{ST} values than long-lived perennials. Long-lived perennials also had significantly lower G_{ST} values than short-lived perennials. This pattern was probably due to the higher proportion of tree species in the long-lived perennial category.

Life form and geographic range

The combination of life form and geographic range explained a low proportion of the among-species variation in P_s and H_{es} and only moderate amounts of variation in G_{ST} values (Table 6.2). The highest values of P_s were associated with long-lived perennial species with regional geographic ranges, while widespread annual species had the highest H_{es} values (Table 6.8). Short-lived, endemic perennials had the lowest P_s and H_{es} values. Annual species with regional distributions had the highest G_{ST} values and long-lived perennials with regional distributions had the lowest G_{ST} values.

Within annual and short-lived perennial life forms geographic range had no discernible effects on P_s, H_{es} and G_{ST} (Table 6.8). Within the long-lived perennial category both P_s and H_{es} increased and G_{ST} decreased significantly with more widely distributed species. Variation in life form had little effect on the levels of genetic diversity contained within endemic species. However, for narrowly and regionally distributed species, genetic diversity was significantly higher in long-lived perennial species. Widespread annual species had more genetic diversity than widespread short-lived perennials. Endemic species with different life forms had different, but non-significant, G_{ST} values due to

Table 6.8. *Mean levels of genetic variation within species and its distribution among populations for combined categories of life form and geographic range*

(See Table 6.3 for definitions of the symbols used.)

Trait combination	n	P_s(%)	H_{es}	G_{ST}
Annual				
endemic	35	50.1[bcd]	0.149[cde]	0.223[cd]
narrow	60	39.6[ef]	0.113[ef]	0.352[b]
regional	65	49.8[cd]	0.143[cde]	0.499[a]
widespread	70	56.5[bcd]	0.200[a]	0.296[bc]
Short-lived perennial				
endemic	30	32.1[f]	0.083[f]	0.325[bd]
narrow	52	49.8[cd]	0.148[cde]	0.216[cd]
regional	93	42.2[def]	0.123[def]	0.280[bc]
widespread	59	48.2[cd]	0.154[cd]	0.194[de]
Long-lived perennial				
endemic	32	48.1[cde]	0.105[ef]	0.150[def]
narrow	70	59.5[ab]	0.163[bc]	0.132[ef]
regional	151	67.0[a]	0.190[ab]	0.086[f]

the low number of entries in these categories. In contrast, for narrow and regionally distributed species, annuals had higher G_{ST} values than short-lived and long-lived perennials. As most long-lived perennials in the database are trees, we conclude that tree species have less genetic differentiation among their populations than other plant species with similar geographic distributions.

Seed dispersal and geographic range

This combination of traits explained a much lower proportion of the variation in G_{ST} and approximately the same levels of variation in P_s and H_{es} as the other combinations (Table 6.2). Widespread species with animal-attached seeds had the highest P_s and H_{es} values, while species with ingested seeds and narrow distributions had the lowest H_{es} and P_s values (Table 6.9). Wind-dispersed species with endemic distributions also had low genetic diversity. Wind-dispersed species with regional distributions had the lowest G_{ST} values, while widespread species with ingested seeds had the highest G_{ST} values.

Within a seed dispersal category, the species' geographic range had little predictable effect on measures of genetic diversity. Species with ingested and

Table 6.9. *Mean levels of genetic variation within species and its distribution among populations for combined categories of seed dispersal mechanism and geographic range*

(See Table 6.3 for definitions of the symbols used.)

Trait combination	n	P_s(%)	H_{es}	G_{ST}
Attached				
narrow	17	68.3[a]	0.214[ab]	0.215[cdef]
regional	45	64.3[ab]	0.192[abc]	0.245[cd]
widespread	31	72.4[a]	0.221[a]	0.201[def]
Gravity				
endemic	53	46.2[d]	0.130[def]	0.198[def]
narrow	85	46.3[d]	0.136[def]	0.291[bc]
regional	126	44.6[d]	0.128[ef]	0.336[ab]
widespread	70	47.8[d]	0.174[bc]	0.247[cd]
Ingested				
narrow	22	32.3[e]	0.097[f]	0.254[bcd]
regional	33	60.1[abc]	0.213[ab]	0.216[cde]
widespread	18	48.6[cd]	0.201[abc]	0.441[a]
Wind				
endemic	34	43.4[de]	0.107[f]	0.259[bcd]
narrow	59	57.2[bc]	0.148[cde]	0.123[ef]
regional	106	63.6[ab]	0.163[bcd]	0.104[f]
widespread	25	55.1[bcd]	0.144[cdef]	0.156[def]

wind-dispersed seeds tended to have significantly more genetic diversity in more widely distributed species, but species with gravity and attached dispersal mechanisms did not demonstrate this pattern. Within the wind-dispersed category endemic species had significantly higher G_{ST} values than either narrow or regionally distributed species, but this pattern was not repeated for the other seed dispersal categories. Within geographic range categories there was no pattern with the seed dispersal mechanisms for P_s. For H_{es}, however, species with attached seed dispersal generally had higher values than species with other seed dispersal mechanisms. Regardless of the geographic range of the species, species with wind-dispersed seeds tended to have lower G_{ST} values than species with other seed dispersal mechanisms. A curious observation is that species with attached seeds generally had high levels of genetic diversity in all of the analyses. This unexpected result may be due to the high number of monocots in this category.

Table 6.10. *Mean levels of genetic variation within species and its distribution among populations for several plant families*

(See Table 6.3 for definitions of the symbols used.)

Family	n	$P_s(\%)$	H_{es}	G_{ST}
Asteraceae	101	45.3	0.127	0.204
Chenopodiaceae	22	40.6	0.099	0.540
Cucurbitaceae	23	40.4	0.168	0.397
Onagraceae	23	34.4	0.106	0.338
Orchidaceae	16	44.8	0.137	0.087
Schrophulariaceae	16	37.2	0.123	0.372
Solanaceae	23	32.0	0.094	0.426
Poaceae	91	62.7	0.201	0.284
Fabaceae	48	59.6	0.184	0.277
Myrtaceae[a]	14	81.8	0.222	0.134
Fagaceae[a]	27	65.3	0.198	0.085
Pinaceae[a]	103	73.0	0.176	0.073

[a] Families with predominantly woody taxa.

3.2 Plant families

The analysis of genetic diversity within plant families indicated that, in general, families with predominantly herbaceous species had less genetic diversity and higher genetic differentiation among their populations than families with predominantly long-lived woody perennials (Table 6.10). There were, however, a few interesting exceptions. In particular, the Orchidaceae had levels of genetic diversity typical of other predominantly herbaceous families but had an exceptionally low mean G_{ST} (Table 6.10). This result was perhaps due to the species-specific pollinators characteristic of orchids, and to their tiny wind-borne seeds. Both of these traits could produce high rates of gene flow among populations. The Poaceae was also atypical for an herbaceous family since its species have high genetic diversity and comparatively less genetic heterogeneity among populations. Families with predominantly woody species, i.e. Fagaceae, Pinaceae and Myrtaceae (= *Eucalyptus* in our data), all had high genetic diversity values and exhibited little differentiation among populations. The only family analysed with a significant mixture of herbaceous and woody taxa was the Fabaceae. Interestingly, its mean genetic diversity parameters were intermediate between the predominantly herbaceous and the predominantly woody families. Comparisons of genetic diversity parameters within the Fabaceae demonstrated that herbaceous legumes ($n = 32$) had mean P_s, H_{es} and G_{ST} values of 53%, 0.160, 0.352 respectively, while values for the woody legumes ($n = 18$) were 76%, 0.229, and 0.124. These results suggest that genetic diversity and its distribution are more closely associated with these individual species' life history traits than with their phylogenetic status.

4 Conclusions

These results reinforce the conclusions of earlier reviews by demonstrating the influence of life history traits on levels and distribution of genetic diversity in seed plants. All traits examined had significant effects on the three genetic diversity parameters considered. Life form and breeding system, in particular, had significant influences on genetic diversity and its distribution. More specifically, outcrossing species have significantly less genetic diversity among their populations, regardless of their other traits. Species with low interpopulation genetic differentiation also tend to have more overall genetic diversity. The most interesting insight arising from the two-trait analyses is the observation that woody plants have lower G_{ST} values than herbaceous plants with the same combinations of life history traits, regardless of their phylogenetic relationship. Earlier reviews (Loveless & Hamrick 1984; Hamrick & Godt 1989) speculated that woody plants had lower G_{ST} values because trees have certain life history traits in common. The present analysis, however, indicates that woody plants have lower G_{ST} values and somewhat higher P_s and H_{es} values than non-woody species that share the same breeding systems, geographic ranges and seed dispersal mechanisms. As discussed earlier, the tall stature and comparatively low population densities of trees should result in greater dispersal distances for pollen and seeds than would occur in populations of shorter, more dense herbaceous species.

The greater potential for gene movement of trees should also affect their ability to maintain genetic diversity (Hamrick & Nason 1996). New alleles arising in populations of outcrossing tree species should have a higher probability of being dispersed into other populations than novel genes introduced into populations of herbaceous species with limited gene dispersal potential. Due to their greater dispersal potential, novel alleles are less likely to be lost to tree species through drift or population extinction. In contrast, novel alleles are more likely to be lost in species that experience less gene flow. These species should, as a result, have fewer polymorphic loci and less overall genetic diversity.

References

Brown, A. H. D. (1979). Enzyme polymorphism in plant populations. *Theoretical Population Biology* **15**, 1–42.

Crawford, D. J. (1983). Phylogenetic and systematic inferences from electrophoretic studies. In *Isozymes in plant breeding*, part A (ed. S. D. Tanksely & J. J. Orton), pp. 257–287. Amsterdam: Elsevier.

Gottlieb, L. D. (1977). Electrophoretic evidence and plant systematics. *Annals of the Missouri Botanical Garden* **64**, 161–180.

Hamrick, J. L. (1978). Genetic variation and longevity. In *Topics in plant population biology* (ed. O. T. Solbrig, S. Jain, G. B. Johnson & P. H. Raven), pp. 84–113. New York: Columbia University Press.

Hamrick, J. L. (1983). The distribution of genetic variation within and among natural plant populations. In *Genetics and conservation* (ed. C. M. Shonewald-Cox, S. M. Chambers, B. MacBryde & L. Thomas), pp. 335–348. Menlo Park, CA: Benjamin/Cummings.

Hamrick, J. L. & Godt, M. J. W. (1989). Allozyme diversity in plant species. In *Plant population genetics, breeding and germplasm resources* (ed. A. H. D. Brown, M. T. Clegg, A. L. Kahler & B. S. Weir), pp. 43–63. Sunderland, MA: Sinauer.

Hamrick, J. L. & Godt, M. J. W. (1996a). Conservation genetics of endemic plant species. In *Conservation genetics: Case histories from nature* (ed. J. C. Avise & J. L. Hamrick), pp. 281–304. New York: Chapman and Hall.

Hamrick, J. L. & Godt, M. J. W. (1996b). Allozyme diversity in cultivated crops. *Crop Science*. (In the press).

Hamrick, J. L., Godt, M. J. W. & Sherman-Broyles, S. L. (1992). Factors influencing levels of genetic diversity in woody plant species. *New Forests* **6**, 95–124.

Hamrick, J. L., Linhart, Y. B. & Mitton, J. B. (1979). Relationships between life history characteristics and electrophoretically detectable genetic variation in plants. *Annual Review of Ecology and Systematics* **10**, 173–200.

Hamrick, J. L. & Nason, J. D. (1996). Consequences of dispersal in plants. In *Population dynamics in ecological space and time* (ed. O. E. Rhodes, R. K. Chesser & M. H. Smith), pp. 203–236. Chicago: University of Chicago Press.

Loveless, M. D. & Hamrick, J. L. (1984). Ecological determinants of genetic structure of plant populations. *Annual Review of Ecology and Systematics* **15**, 65–95.

Nei, M. (1973). Analysis of gene diversity in subdivided populations. *Proceedings of the National Academy of Sciences of the U.S.A.* **70**, 3321–3323.

SAS Institute Inc. (1988). *SAS/Stat User's Guide*, Release 6.03 Edition. Cary, NC: SAS Institute, Inc.

Weir, B. S. (1990). *Genetic data analysis*. Sunderland, MA: Sinauer.

III · Seeds

7 · Evolutionary ecology of seed dormancy and seed size

Mark Rees

1 Introduction

Seeds provide an essential link in population dynamics by allowing the establishment of new individuals, and so the founding of populations. Patterns of seed dispersal, both in time and space, determine who interacts with whom, and hence the strength of interactions both within and between species, which in turn determines observed patterns of species diversity (Law & Watkinson 1989; Tilman & Pacala 1993; Rees *et al.* 1996). Therefore, when considering the evolutionary biology of seed traits it is essential to place this in its ecological context. Likewise when considering the ecology of seed traits we must also recognise their evolutionary significance. For example, theoretical models have demonstrated that seed dormancy can play an important role in allowing coexistence via the 'storage effect' (Chesson 1988). This occurs when there is temporal variation in the environment, and the presence of long-lived seeds allows different species to be favoured at different times. Clearly, this coexistence mechanism is only likely to be important in those communities where dormancy has evolved, such as in desert annuals (Venable *et al.* 1993) or weeds of arable fields (Rees & Long 1992). However, in other communities (for example those consisting of sand dune annuals), coexistence via a temporal 'storage effect' is unlikely because there are microsites available for colonization in virtually every year, so evolution favours complete germination, with the result that few species have dormant seeds (Kelly 1982; Watkinson & Davy 1985).

Clearly, seed traits are embedded in a plant life cycle and so changes in the traits of established plants will have fitness consequences for seed traits and vice versa (Salisbury 1942, 1974; Mazer 1989; Rees 1993, 1994; but see Grime 1979, Grime *et al.* 1987). This means we should expect to find correlations between seed and established plant traits, which is indeed the case (Salisbury 1942; Silvertown 1981; Mazer 1989; Rees 1993), but also means that we need to be careful when analysing comparative data because many traits are likely to covary simultaneously, making it important to include several traits and phylogenetic information into analyses. Simply treating species as independent data points in statistical analyses can obscure important ecological and evolutionary relationships (Harvey & Pagel 1991; Garnier 1992; Harvey *et al.* 1995; Rees 1995).

In the rest of the chapter I first describe why seed dormancy presents a problem in evolutionary biology, and then review the various theoretical mechanisms that could explain the evolution of dormancy. For each of these mechanisms I outline appropriate comparative tests and present some relevant data analysis. I then briefly describe the theory, such as it is, for the evolution of seed mass, and present some comparative tests. General conclusions from the comparative analyses presented are discussed.

2 Why have dormant seeds?

Why is dormancy a problem? The general answer is that we would normally expect individual plants to reproduce as soon as they could. There are two main reasons for this:

1. In an increasing population, early reproduction is favoured because of the multiplicative nature of population growth.
2. Individuals that forgo reproduction may die before they are able to reproduce.

Both processes impose a cost of delayed reproduction (Bulmer 1985). Hence, from an evolutionary perspective, seed dormancy, which is a form of delayed reproduction, presents a problem; this has led evolutionary biologists to explore the conditions under which the evolution of seed dormancy might be favoured.

It is instructive to begin by looking carefully at the assumptions of models which suggest that there is a cost to remaining dormant. The models assume:

1. Seeds have no information on the quality of the environment.
2. Competition occurs primarily between individuals that are not related rather than between siblings.
3. There is no temporal variation in the quality of the environment.

Allowing seeds to detect the quality of the environment will favour the evolution of dormancy even in a temporally constant environment, providing there is spatial variation in establishment conditions (de Jong *et al.* 1987).

If competition occurs primarily between sibs, as a result of dispersal inside multi-seeded fruits, then dormancy may be favoured in a temporally constant environment, even when seeds have no information on the quality of the environment (Ellner 1986). This is a result of a parent–offspring conflict which the parent wins by using the seed coat to prevent the embryo from germinating. The conflict occurs because parents wish to reduce competition between offspring, whereas the individual offspring whose germination is delayed have lower inclusive fitness than those which germinate. Thus, selection on embryos favours complete germination, whereas selection on the parent favours delayed germination for some of the offspring.

Allowing the environment to vary from one year to the next can also select for the evolution of dormancy even when seeds have no information on the quality of the environment and sib competition is weak. Consider the case where the conditions are suitable for growth and reproduction in some years, but reproduction fails completely in other years. In such an environment an annual plant genotype with no seed dormancy would maximise its arithmetic average population growth rate (λ) but would become extinct the first time that reproduction failed completely. At the other extreme, a genotype that never germinated would also become extinct as a result of seed mortality. Hence, in a variable environment we would expect an intermediate germination strategy to be optimal (Cohen 1966; Bulmer 1984; Ellner 1987).

Theory also predicts that life history attributes that reduce the impact of environmental variation on fitness will show patterns of negative covariation (Venable & Brown 1988; Rees 1993, 1994). For example, species with efficient dispersal in space, either as a result of seed dispersal or vegetative growth, reduce the likelihood that all seeds will be exposed to unfavourable conditions in any one year, and so we would expect a negative relationship between the efficiency of spatial dispersal and dormancy. It has also been suggested that a trade-off between spatial and temporal seed dispersal might arise as a result of physical and biochemical constraints (Lokesha et al. 1992). These authors argue that packing seeds with fats allows seed mass to be reduced while maintaining energy content because fats contain more energy per unit mass than proteins or carbohydrates. Hence in wind-dispersed species, where the efficiency of dispersal depends, in part, on seed mass because heavy seeds fall more rapidly than light ones, seeds should contain a higher proportion of fat than protein or carbohydrate. However, the use of fats has several disadvantages: their synthesis is more energy-demanding than the production of proteins or carbohydrates; and lipid autoxidation is thought to cause the disruption of several cell components resulting in loss of viability (Ponquett

Table 7.1. *Theoretical patterns of trait covariation with seed dormancy.*

(↑ indicates that dormancy is expected to increase with an increase in the trait; ↓ indicates that dormancy is expected to decrease with an increase in the trait. Note that in several cases the competing theories make similar predictions making it difficult to distinguish between them using comparative tests.)

Theory	Traits					
	Adult longevity	Adult lateral spread	Adult height	Spatial dispersal	Seed death rate	Seed mass
Bet-hedging	↓,↑	→			→	↓,↑
Sib-competition				→	→	
Cueing				→	→	
Chemical packing			→	→	→	

et al. 1992). Therefore, species with wind-dispersed seeds should have a high fat content, and as a result the seeds are short-lived, whereas species that are not wind-dispersed will have a lower fat content and so have greater seed longevity. There is some evidence that wind-dispersed seeds do indeed contain a higher proportion of fat than seeds that are passively dispersed or dispersed by animals (Lokesha *et al.* 1992). Species with large individual seeds are predicted to have reduced dormancy because their seedlings can draw on a larger food reserve, and hence establish in relatively unfavourable environments. In a similar way species with long-lived adults are buffered from temporal variation in the environment and this also selects for less dormancy (Venable & Brown 1988; Rees 1994). A summary of the various theoretical predictions is given in Table 7.1. Some of these ideas were tested using modern comparative methods, with taxonomy as a surrogate for a phylogeny, by Rees (1993), who found that large-seeded species do have less dormant seeds; species with efficient seed dispersal in space have less seed dormancy; and long-lived species also have less dormancy.

3 Data and comparative methods – seed dormancy

Two datasets were used in the analyses, the first was collected by Harold Roberts (Roberts 1964, 1979; Roberts & Boddrell 1983), the second is the Sheffield database (Grime *et al.* 1988, Hodgson *et al.* 1995). Roberts's data allows a quantitative estimate of seed dormancy, (see below), whereas in the Sheffield database species are categorized on a three point scale: type 1 transient, seeds rarely persisting for more than one year; type 2 short-term persistent, seeds persisting for more than one year but usually less than five; and type 3, long-term persistent seed persisting for at least five years, and often much longer (Hodgson *et al.* 1995).

In Roberts's experiments, freshly collected seeds were mixed with the upper 7.5 cm of steam-sterilized sandy clay loam confined in open-ended 23 cm diameter cylinders sunk in the ground outdoors and netted to exclude birds. There were two replicate cylinders per species, each containing usually 1000 seeds. On three occasions each year the soil layer containing the seeds was mixed to its full depth to simulate cultivation. All seedlings were recorded and removed soon after they appeared. Recording continued usually for five years; the soil dug up and the number of viable seeds remaining determined. For most species, separate experiments were begun in each of three different years using freshly collected seeds. Between 1953 and 1986 data were obtained for 171 species. A species list for the annual and perennial forbs is given in Rees & Long (1993).

Hence, for each species we have a pattern of seedling emergence times describing the number of recruits in each year of the experiment. The main variable analysed was the number of seedlings that emerge during the first year divided by the total number of seedlings that emerge during the 5 years of the experiment; this quantity was termed p1. Note that because seed mortality is completely unobserved, this quantity cannot be simply interpreted as a probability of germination (see Rees & Long (1993) for a discussion of this point). However, species with high germination rates would be expected to have high p1 values.

Species with high seed mortality will automatically have higher p1 values than species with low seed mortality, simply because those seedlings that emerge in year 1 will be a greater proportion of those that ever emerge. However, analyses using the proportion of all seeds sown that emerge in year 1, called s1, gave similar results: s1 will be sensitive to differences in initial seed viability but relatively insensitive to differences in seed mortality during the experiment. Hence, the results presented largely reflect differences in seed germination probability.

The information on p1 was integrated into a database which contained life history information on each of the species. The variables used and the data sources were:

1. Adult longevity was determined using a simple scoring system, annual = 1, biennial = 2, monocarpic = 3, perennial = 5. Species in composite categories were given a weighted score; the first term given a weight of 1, the second 0.5 (e.g. annual/perennial = (1 + 0.5*5)/(1 + 0.5)). (Data sources: Grime *et al.* 1988; Clapham *et al.* 1989; Hodgson *et al.* 1995.)
2. Adult lateral spread and adult height were coded into categories following Hodgson *et al.* (1995). Where possible the species were coded using the information in Hodgson *et al.* (1995), species not in this database were classified using information from Clapham *et al.* (1989) and Stace (1991).
3. Seed masses (mg) were obtained from Grime *et al.* (1988) and also the ecological flora database (Fitter & Peat 1994). Seed masses were log transformed prior to analysis.
4. Spatial seed dispersal was assessed using the data in Hodgson *et al.* (1995).

The relationships between these traits, p1 and the Sheffield three-point seed bank score were explored using independent contrasts, calculated using the CAIC package (see Purvis & Rambaut (1995) for details). The phylogeny used in the calculation of the contrasts was phylogeny B of Chase *et al.* (1993). A detailed discussion of the rational underlying this method of analysis is given in Harvey & Pagel (1991).

4 Results – seed dormancy

4.1 Relationships between established plant traits

Considering the species in Roberts's dataset first we find that the relationship between adult longevity and lateral spread was positive and statistically significant ($r^2 = 0.79$, $p < 0.0001$, $n = 41$). Surprisingly there was no significant relationship between lateral spread and plant height ($r^2 = 0.005$, $p > 0.05$, $n = 41$) or between longevity and height ($r^2 = 0.003$, $p > 0.05$, $n = 41$). Multiple regression analyses, modelling variation in the contrast for one variable as a linear function of the other two, gave similar results. Analyses using species as independent data points also gave qualitatively similar results: there were no significant relationships between height and either lateral spread or longevity. However the relationship between lateral spread and longevity was highly significant ($r^2 = 0.53$, $p > 0.0001$, $n = 169$).

In the Sheffield dataset we find similar results: the relationship between adult longevity and lateral spread is positive and statistically significant ($r^2 = 0.46$, $p < 0.0001$, $n = 113$). The relationship between height and adult longevity is not statistically significant ($r^2 = 0.03$, $p > 0.05$, $n = 113$), but in contrast to Roberts's data set there was a positive, statistically significant relationship between height and lateral spread ($r^2 = 0.15$, $p < 0.001$, $n = 113$). These results are mirrored in analyses that use species as independent data points; the relationship between adult longevity and lateral spread was positive and highly significant ($r^2 = 0.48$, $p < 0.0001$, $n = 513$), the relationship between plant height and adult longevity was not significant ($r^2 = 0.01$, $p > 0.05$, $n = 513$), however there was a positive, statistically significant relationship between height and lateral spread ($r^2 = 0.10$, $p < 0.0001$, $n = 513$).

4.2 Relationships between p1, seed bank scores and other traits

In the p1 analyses plant height was never significant in any regression model. However, there were highly significant positive relationships between lateral spread, seed mass, adult longevity and p1 (p1 *vs* lateral spread $r^2 = 0.27$, $p < 0.001$, $n = 41$, p1 *vs* seed mass $r^2 = 0.30$, $p < 0.001$, $n = 41$, p1 *vs* longevity $r^2 = 0.49$, $p < 0.001$, $n = 41$; see Figure 7.1). In a multiple regression

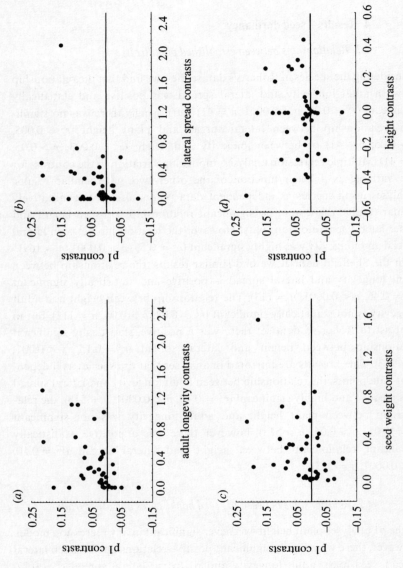

Figure 7.1. Independent contrast analyses of p1, the proportion of all seedling observed that emerge in year 1. Relationships between p1 and, (a) adult longevity, (b) lateral spread, (c) seed weight, and (d) height. See text for details.

model for p1 with all four variables, only seed mass and adult longevity were statistically significant (seed mass $p < 0.001$, longevity $p < 0.001$, $r^2 = 0.65$, $n = 41$).

Analysing only those species that have no mechanisms for spatial seed dispersal (i.e. those with no morphological features to aid dispersal, classified by Hodgson *et al.* (1995) as unspecialised, plus those dispersed by ants), we find that neither seed mass nor lateral spread entered the analysis significantly, either when alone or with the height and/or adult longevity ($p < 0.05$ in all cases). In contrast adult longevity was significant, both when entered alone, and also with the other variables (longevity alone, $r^2 = 0.36$, $p < 0.001$, $n = 29$; longevity with seed mass, height and lateral spread, $r^2 = 0.50$, $p < 0.001$, $n = 29$). When considering species with efficient mechanisms for spatial dispersal (i.e. those with wind- or animal-dispersed seeds), we find no significant relationships between p1 and seed mass or plant height, but there were positive statistically significant relationships between p1 and lateral spread and between p1 and adult longevity (lateral spread $r^2 = 0.49$, $p < 0.01$, $n = 16$; adult longevity $r^2 = 0.63$, $p < 0.001$, $n = 16$). In the full model with adult longevity, lateral spread, seed mass and height, only adult longevity was significant ($r^2 = 0.65$, $p < 0.05$, $n = 16$).

Analyses of the complete Sheffield dataset demonstrate that lateral spread, seed mass and adult longevity were all negatively related to seed bank score; there was no significant effect of plant height. This indicated that large-seeded plants, long-lived plants and those with wide lateral spread are less likely to have long-lived seed banks, in agreement with the analysis of Roberts's dataset. However, in contrast to Roberts's dataset, in the full model adult longevity was not statistically significant; only seed mass and lateral spread entered the full model significantly, see Table 7.2).

The dataset was then divided into three dispersal categories; unspecialized, wind-dispersed and animal-dispersed, using the data in Hodgson *et al.*(1995). In the set of species with seeds unspecialized for spatial dispersal there were negative relationships between lateral spread and seed bank scores, and also between adult longevity and seed bank scores, but there was no relationship with seed mass or height, (see Table 7.2). Species with wind- or animal-dispersed seeds show similar relationships between seed bank scores and lateral spread, longevity and seed mass when these variables are used as the only explanatory variable in an analysis. However, in the full model wind-dispersed seeds show a negative relationship between lateral spread, adult longevity and seed bank scores, and there is a marginally significant positive relationship between seed bank scores and plant height, (see Table 7.2). In

Table 7.2. *Regression models for seed bank score contrasts*

(In the single effect models only a single term was included in each model, in the full model all four terms were included simultaneously. The table entries give the sign of the relationship, significance levels (n.s. $= p > 0.05$, $* = p < 0.05$, $** = p < 0.01$, $*** = p < 0.001$), and for the single variable models the coefficient of determination r^2. The coefficient of determination for the full model and the number of contrasts in each of the analyses are given in the right-hand column. All regressions were forced through the origin.)

	Single variable model				Full model				r^2 and n for full model
	Adult height	Lateral spread	Seed mass	Adult longevity	Adult height	Lateral spread	Seed mass	Adult longevity	
Full data set	n.s. $r^2 = 0.002$	$-***$ $r^2 = 0.16$	$-***$ $r^2 = 0.20$	$-***$ $r^2 = 0.15$	n.s.	$-**$	$-**$	n.s.	$r^2 = 0.35$ $n = 106$
Unspecialised	n.s. $r^2 = 0.08$	$-*$ $r^2 = 0.14$	n.s. $r^2 = 0.003$	$-**$ $r^2 = 0.20$	n.s.	n.s.	n.s.	n.s.	$r^2 = 0.23$ $n = 40$
Wind dispersed	n.s. $r^2 = 0.006$	$-***$ $r^2 = 0.33$	$-**$ $r^2 = 0.21$	$-**$ $r^2 = 0.22$	$+*$	$-***$	$-**$	n.s.	$r^2 = 0.60$ $n = 30$
Animal-dispersed	n.s. $r^2 = 0.01$	$-$ $r^2 = 0.08$	$-***$ $r^2 = 0.60$	$-*$ $r^2 = 0.09$	n.s.	n.s.	$-***$	n.s.	$r^2 = 0.65$ $n = 47$

those species dispersed by animals, only seed mass was significant in the full model.

5 Evolution of seed mass

Salisbury (1942) argued that seed mass reflected the establishment conditions that seedlings experienced. More precisely he argued 'that the large seed, with its copious provision of food, will be especially advantageous in closed communities where the colonising individual must be capable of growing above the surrounding vegetation, or at least into a level of moderate illumination, before it can receive a sufficient intensity of light to manufacture its own food at a rate comparable to that of its neighbours.' Tilman developed these ideas within the context of his size-structured model of competition called Allocate, and came to similar conclusions to Salisbury (Tilman 1988): namely that increased seed mass would be favoured in habitats where competition for light is important. Tilman also noted that increased seed mass would be favoured in habitats with resource-poor soils if resource availability increased with depth. In Tilman's model all species have identical adult traits (i.e. allocation patterns to roots, shoots and photosynthetic tissues) and exclusion of small-seeded species occurs as a result of competition between juveniles. These ideas lead to two predictions: (1) tall species should have larger seeds than short species; (2) perennials should have larger seeds than annuals. These predictions assume that as plants get taller, competition for light becomes more important, and that perennials typically occur in closed habitats where competition for light is more intense.

The standard treatment of the evolution of offspring size assumes a simple relationship between investment per offspring and offspring success (Smith & Fretwell 1974; Lloyd 1987), and uses this assumed relationship to derive ESS (evolutionarily stable strategy) or optimality solutions. However, by assuming the existence of a particular relationship between investment in each offspring and its subsequent success, we rule out the possibility that frequency-dependent processes affect fitness (Lloyd 1987). This severely restricts the range of ecological processes that could determine the relative fitness of different sized offspring. For example, if we assume seed mass determines competitive ability, which in turn affects fitness, then to evaluate fitness we would need to know the distribution of seed masses that an individual competes against (Parker & Begon 1986): a 2-mg seed might be a good competitor against 1-mg seeds but a poor one against 3-mg seeds. This means that the curves used in the standard Smith–Fretwell analysis are not fixed but depend on the frequency distribution of seed sizes in a population or community. Therefore to study traits that affect competitive ability we require a game theory approach which allows the analysis of frequency-dependent coevolutionary games (e.g. Brown & Vincent 1987; Vincent *et al.* 1993).

The standard analysis, with its fixed relationship between resources invested and fitness, predicts a single evolutionary optimum investment per offspring (Smith & Fretwell 1974; Lloyd 1987). However, even within guilds within communities there is often a wide range of seed masses; for example in the four-species guild of sand dune annuals studied by Rees *et al.* (1966) there was a 27-fold variation is seed mass. Hence, we need to reconcile the existence of broad variation in seed mass within guilds and communities with the theoretical prediction of a single evolutionary optimal seed mass. One obvious explanation for the wide range of seed masses observed is that establishment conditions vary enormously within communities (see below).

Recent theoretical work has extended the classical frequency-independent models to include frequency dependence (Geritz 1995; Rees & Westoby 1997). These models assume seed mass determines the number of progeny produced per unit reproductive effort (i.e. there is a size number trade-off), and competitive ability. The outcome of competition is assumed to be determined by seed mass. Competition between individuals derived from seeds of different masses is assumed to be asymmetric, with individuals from larger seeds having a greater negative effect on individuals derived from small seeds than vice versa. In Geritz's models a single seed mass is never evolutionarily stable, whereas in the models of Rees & Westoby a single seed mass can be evolutionarily stable but generally the ESS is made up of several species/strategies each with a different seed mass (Geritz 1995; Rees & Westoby 1997). Hence, in plant communities we should expect to see a range of seed masses, and this is indeed the case (Salisbury 1942). It is important to note that one of the key assumptions of Geritz's model, which results in many seed masses being present in the ESS, is that different species/strategies generate small-scale spatial variation in establishment conditions. Large-seeded species/strategies exclude small-seeded ones from microsites as a result of shading or nutrient uptake, but large-seeded species/strategies produce fewer seeds and so have a disadvantage in colonization. Hence, the new models can be thought of as spatial extensions of Salisbury's original idea, with the presence or absence of different species/strategies generating the variation in establishment conditions. An important corollary of the coexistence of many seed masses within the models is that a component of seed size variation may not be correlated with other life history traits (Rees & Westoby 1997).

However, these models do not tell us how seed mass should vary in relation to plant height, lateral spread or longevity. Geritz has produced a model for perennial species which has predictions similar to the annual models, but in the perennial model the maximum seed mass is set by the yearly resource allocation to reproduction, termed R_i. To predict how maximum seed mass varies with perenniality we need to know how R_i varies with plant longevity. Long-lived plants generally have a higher percentage allocation to structural, non-reproductive structures than annuals; whereas annuals in contrast have higher percentage allocation to reproduction than perennials (Tilman 1988; Wilson & Thompson 1989; Silvertown & Dodd this volume). Hence, if annuals and perennials have the same total mass, annuals will have a larger R_i and hence should produce larger seeds, on average. If, however, perennial plants are larger then this could offset the lower percentage allocation to reproduction with the result that perennials have a larger R_i than annuals, which gives the opposite prediction. Therefore, in order to test these ideas we need to know more about R_i and adult longevity.

It has been suggested that plant height might alter the efficiency of seed dispersal, particularly in species with unassisted or wind-dispersed seeds, because small seeds travel further than large ones for a given height of release (Rabinowitz & Rapp 1981; Thompson & Rabinowitz 1989). Hence, short plants might be constrained to have small seeds, whereas tall plants can potentially have large seeds. A corollary of this is that in species dispersed by animals there should be no relationship between plant height and seed mass. The difficulty with these hypotheses is that they are framed in terms of dispersal efficiency and it may not be straightforward to link this with fitness.

A final set of hypotheses link seed mass with mechanisms that allow escape from predators or competition (Janzen 1970; Eriksson 1992). Several hypotheses can be developed. For example, in species with seeds unspecialized for spatial dispersal we might expect those with extensive clonal spread to produce small seeds. There are two potential advantages of this: (1) average dispersal distance might be increased; and (2) more small seeds are produced, which increases the absolute number that disperse a given distance. In each case competition between seedlings and clonal ramets will be decreased, which could increase fitness. A corollary of this is that no decrease in seed mass with clonal extent should be observed in species with wind- or animal-dispersed seeds.

Table 7.3. *Regression models for log seed mass contrasts*

(In the single effect models only a single term was included in each model, in the full model all three terms were included simultaneously. The table entries give the sign of the relationship, significance levels (n.s. = $p > 0.05$, * = $p < 0.05$, ** = $p < 0.01$, *** = $p < 0.001$), and, for the single variable models, the coefficient of determination r^2. The coefficient of determination for the full model and the number of contrasts in each of the analyses are given in the right-hand column. All regressions were forced through the origin. In those species with seeds dispersed by animals, two sets of figures are given; those in brackets refer to a reduced data set where a single outlier was excluded. This outlier was a comparison between *Viburnum opulus* and *Sambucus nigra*.)

	Single variable model			Full model			r^2 and n for full model
	Adult height	Lateral spread	Adult longevity	Adult height	Lateral spread	Adult longevity	
Full dataset	+** $r^2 = 0.07$	+*** $r^2 = 0.10$	n.s. $r^2 = 0.02$	n.s.	+*	n.s.	$r^2 = 0.12$ $n = 110$
Unspecialised	+* $r^2 = 0.15$	n.s. $r^2 = 0.02$	n.s. $r^2 = 0.05$	+***	−**	+***	$r^2 = 0.45$ $n = 42$
Wind-dispersed	−*** $r^2 = 0.45$	n.s. $r^2 = 0.01$	+*** $r^2 = 0.48$	+**	n.s.	+***	$r^2 = 0.65$ $n = 33$
Animal-dispersed	n.s. $r^2 = 0.06$	+* $r^2 = 0.10$	n.s. $r^2 = 0.03$	n.s.	n.s.	n.s.	$r^2 = 0.11$ $n = 48$
	(+** $r^2 = 0.18$)	(+** $r^2 = 0.14$)	(n.s. $r^2 = 0.04$)	(n.s.)	(n.s.)	(n.s.)	($r = 0.21$) $n = 47$

6 Data and comparative methods – seed mass

The analyses use data from the Sheffield flora (Grime *et al.* 1988; Hodgson *et al.* 1995). Seed masses were taken from Grime *et al.* (1988) and from the ecological flora database (Fitter & Peat 1994); the resulting dataset contained information on seed masses for 382 species. Species were classified into three dispersal strategies (i.e. unspecialized, wind-dispersed and animal-dispersed) using Hodgson *et al.* (1995). Details of the data coding are given in Hodgson *et al.* (1995) and also section 3. The relationships between height, lateral spread, adult longevity and seed mass were explored using independent contrasts, calculated using the CAIC package (see Purvis & Rambaut 1995 for details). The phylogeny used in the calculation of the contrasts was phylogeny B of Chase *et al.* (1993).

7 Results – seed mass

In the full dataset there were positive, highly significant relationships between seed mass and plant height and also between seed mass and lateral spread. However, in the full model only adult lateral spread entered significantly. There was no significant relationship between seed mass and adult longevity (Table 7.3). Restricting the analysis to species with seeds unspecialized for dispersal we find that adult height, lateral spread and longevity all enter the multiple regression model significantly; having corrected for height and longevity the relationship between seed mass and lateral spread was negative. Within the wind-dispersed species lateral spread does not enter the regression model significantly but both height and longevity show positive significant relationships. In those species that are animal dispersed, lateral spread is significant when considered in isolation, but in the full model no single factor was significant (see Table 7.3). Excluding a single data point resulted in height being significant in the single variable model (Table 7.3).

An alternative approach to dividing the dataset into three dispersal categories is to include the effects of wind- and animal-dispersal using dummy variables. To do this we need to code two variables: one for wind dispersal that contains a 1 if the species has wind-dispersed seeds and a 0 otherwise. In a similar fashion a variable was coded for animal dispersal. In effect this allows different intercepts to be fitted in the regression models. The full model contained six explanatory variables: adult longevity, height, lateral spread, wind dispersal, animal dispersal, and seed bank scores. The resulting regression model had the following form,

seed weight contrast = 0.36*** height − 0.48**
lateral spread + 0.32* longevity − 0.56*** seed
bank − 0.07 wind + 0.20* animal

where the superscripts indicate the significance levels, (see table legend to Table 7.3 for details), overall $r^2 = 0.44$, $n = 93$. In this multiple regression model only the effect of wind dispersal was not significant. Seed mass was positively related to plant height, longevity, and animal dispersal, and negatively related to lateral spread and seed bank score. Hence, tall plants, long-lived plants and those that are dispersed by animals tend to have heavy seeds, whereas species with extensive lateral spread or that have long-lived seed banks have lighter ones.

8 Discussion

The results of the comparative analysis of p1 and seed bank scores strongly support the predictions of simple bet-hedging theories (Venable & Brown 1988; Rees 1993, 1994). Evidence for spatial and temporal risk-spreading strategies being negatively correlated with dormancy is found for both adult longevity and lateral spread. The relationships between p1, seed bank scores and seed mass are more difficult to interpret because increased seed mass is linked with both increased establishment success, which reduces expected dormancy levels (Venable & Brown 1988), and also with increased levels of predation (Thompson 1987). Increased predation levels make the cost of forming a seed bank greater, which selects for less seed dormancy. In this respect it is perhaps significant that the clearest relationship between dormancy and seed mass occurs in animal-dispersed species.

The impact of spatial seed dispersal was difficult to assess in the current datasets because dispersal traits are relatively deeply rooted in the phylogenetic tree, resulting in few contrasts (for Roberts's data 171 species yielded only eight contrasts, seven of which were positive, suggesting that species with efficient seed spatial dispersal mechanisms have less dormancy, $p = 0.075$). Analysis of the Sheffield dataset was equally uninformative as this problem is compounded with the three-point seed bank score, which results in many uninformative, zero contrasts.

The analyses of seed mass are particularly interesting, suggesting an interplay between dispersal limitation, escaping competition with clonal ramets, and differences in establishment success linked with plant height and longevity. As predicted by the escape hypotheses, the only negative relationship between seed mass and lateral spread was found in those species with seeds unspecialized for dispersal (Eriksson 1992). In the full model (see Table 7.3) plant height was most strongly related to seed mass in those species with seeds unspecialized for spatial dispersal or wind dispersed, in agreement with Thompson and Rabinowitz's ideas on dispersal limitation (Thompson & Rabinowitz 1989).

The relationship between adult longevity and seed mass in species with seeds unspecialized for spatial dispersal, and those dispersed by wind, is consistent with Salisbury's prediction that seed mass should increase in closed habitats (Salisbury 1942, 1974). Differences between open and closed habitats are often confounded with changes in plant longevity and height. For example, Salisbury (1942, 1974) showed, by comparing congeneric species, that species from open (early successional) habitats typically had lighter seeds than those from closed (later successional) habitats. However, in Salisbury's study, 63% of the species from open habitats were annuals compared with only 2% from the closed habitats. Clearly, adult longevity, height and habitat will often be confounded and this will make interpretation of comparative analyses difficult. Why there is no relationship between adult longevity and seed mass in animal-dispersed species is unclear.

The coefficients of determination (r^2) for the multiple regression models of seed mass range from 0.11 to 0.65. Might differences in establishment conditions be the primary source of this unaccounted variation? Recently Leishman et al. (1995) have argued that this is not the case. Their argument rests on two observations: (1) in five floras the variation in seed mass between species is large relative to variation between floras (Leishman et al. 1995), and (2) studies that include habitats (e.g. Mazer 1989) find that habitats account for a relatively small proportion of the variance. However, characterizing the establishment conditions within a flora using a single number is clearly an extremely crude form of analysis. For example, the Sheffield flora used in Leishman et al.'s analysis contains habitats ranging from scree slopes to mires and deciduous woodland. Within each of these habitats there will be enormous variation in establishment conditions. For example, woodlands contain shaded areas and gaps, and within each of these there will be places with ground cover, some with deep litter and some with shallow litter. It is therefore not surprising, given the enormous variation in establishment conditions within each flora, that differences between floras account for a small proportion of the variance in seed mass. In addition to this general argument it should be noted that if the success of a particular seed mass in exploiting any given set of establishment conditions depends on the seed masses of the

Table 7.4. *Regression analyses of independent contrasts for seed mass and height, structured by habitat*

(Each species was classified using its commonest terminal habitat into one of the seven primary habitats; see Hodgson *et al.* 1995 for a detailed description of the habitat classifications.)

Habitat	r^2	n	p
Skeletal	0.64	26	< 0.0001
Grasslands	0.01	29	n.s.
Spoil	0.07	23	n.s.
Wastelands	0.17	21	0.06
Woods	0.01	36	n.s.
Wetlands	0.10	41	0.03
Arable	0.46	20	< 0.001

other species present then the interpretation of comparisons between floras becomes even more problematic.

Similar arguments can be put forward for why habitat variables often account for a relatively small proportion of the variance in seed mass (but see Hammond & Brown (1995), where gap environment assessed on a three-point scale accounted for 31% of the variance in seed mass). It also appears from preliminary analysis of the Sheffield flora that the relationship between seed mass and height may require the fitting of interaction terms (i.e. different intercepts *and* slopes within each habitat). For example, in the seven primary habitats (arable, wetland, skeletal, grassland, spoil, wasteland and woodland) of the Sheffield flora the coefficients of determination (r^2) for the relationship between seed mass and height range from < 0.01 to 0.64, strongly suggesting the possibility of an interaction between habitat and height (Table 7.4). Ignoring interaction terms will reduce the explanatory power of both height and habitat in this case.

A common problem with many studies that compare the use of different variables in predicting say, seed mass, is that the variables are often measured on very different scales. For example, height might be measured on a six point continuous scale whereas longevity might be assessed as annual or perennial. Obviously perenniality will be a much cruder predictor of seed mass than plant height, and we should not be surprised if longevity accounts for a relatively small proportion of the variance in seed mass, which is generally the case (Leishman & Westoby 1994; Leishman *et al.* 1995). Performing a simulation, where the scale of measurement can be changed, clearly illustrates this effect. Assuming we can accurately estimate adult longevity we find, for the simulated data, that the coefficient of determination (r^2) between seed mass and adult longevity is 0.98; however, if longevity is assessed on a two-point scale, annual and perennial, the coefficient of determination drops to 0.38. Interestingly, in one of the few studies where plant size has been accurately quantified (Wilson & Thompson 1989), the relationship between log seed weight and total plant biomass for annual grass species has a coefficient of determination of 0.97 ($n = 8$, using species as independent data points). Likewise, studies which incorporate habitat variables often do not quantify variation in establishment conditions within habitats (but see Mazer (1989), where a four-point scale was used) and so weak descriptive power should expected.

More theoretical work is required to explore the evolution of seed traits in realistic models. Current models either ignore space completely or assume large-scale dispersal so that the positions of sites relative to one another can be ignored. Likewise, most models assume that all plants are the same size, and so make no predictions about how traits should vary in relation to lateral spread or height. Obviously size-dependent spatial models will be difficult to analyse, but have the great advantage that they make predictions about quantities that are easily measured. Coupling careful comparative analyses with modelling studies offers the possibility of rapidly advancing our understanding of the selective forces shaping plant life-histories.

References

Brown, J. S. & Vincent, T. J. (1987). Coevolution as an evolutionary game. *Evolution* **41**, 66–79.

Bulmer, M. G. (1984). Delayed germination of seeds: Cohen's model revisited. *Theoretical Population Biology* **26**, 367–377.

Bulmer, M. G. (1985). Selection for iteroparity in a variable environment. *American Naturalist* **126**, 63–71.

Chase, M. W. *et al.* (1993). Phylogenetics of seed plants: an analysis of nucleotide sequences from the plastid gene rbcL. *Annals of the Missouri Botanical Garden* **80**, 528–580.

Chesson, P. L. (1988). Interactions between environment and competition: how fluctuations mediate coexistence and competitive exclusion. In *Community ecology. Lecture notes in biomathematics* (ed. A. Hastings), pp. 51–71. London: Springer-Verlag.

Clapham, A. R., Tutin, T. G. & Moore, D. M. (1989). *Flora of the British Isles.* Cambridge University Press.

Cohen, D. (1966). Optimising reproduction in a randomly varying environment. *Journal of Theoretical Biology* **12**, 110–129.

de Jong, T. J., Klinkhamer, P. G. L. & Metz, J. A. J. (1987). Selection for biennial life histories in plants. *Vegetatio* **70**, 149–156.

Ellner, S. (1986). Germination dimorphisms and parent-offspring conflict in seed germination. *Journal of Theoretical Biology* **123**, 173–185.

Ellner, S. (1987). Competition and dormancy: a reanalysis and review. *American Naturalist* **130**, 798–803.

Eriksson, O. (1992). Evolution of seed dispersal and recruitment in clonal plants. *Oikos* **63**, 439–448.

Fitter, A. H. & Peat, H. J. (1994). The ecological flora database. *Journal of Ecology* **82**, 415–425.

Garnier, E. (1992). Growth analysis of congeneric annual and perennial grass species. *Journal of Ecology* **80**, 665–675.

Geritz, S. A. H. (1995). Evolutionarily stable seed polymorphism and small-scale spatial variation in seedling density. *American Naturalist* **146**, 685–707.

Grime, J. P. (1979). *Plant strategies and vegetation processes.* Chichester: Wiley.

Grime, J. P., Hunt, R. & Krzanowski, W. J. (1987). Evolutionary physiological ecology of plants. In *Evolutionary physiological ecology* (ed. P. Calow), pp. 105–125. Cambridge University Press.

Grime, J. P., Hodgson, J. G. & Hunt, R. (1988). *Comparative plant ecology.* London: Unwin Hyman.

Hammond, D. S. & Brown, V. K. (1995). Seed size of woody plants in relation to disturbance, dispersal, soil type in wet neotropical forests. *Ecology* **76**, 2544–2561.

Harvey, P. H. & Pagel, M. D. (1991). *The comparative method in evolutionary biology.* Oxford University Press.

Harvey, P. H., Read, A. F. & Nee, S. (1995). Why ecologists need to be phylogenetically challenged. *Journal of Ecology* **83**, 535–536.

Hodgson, J. G., Grime, J. P., Hunt, R. & Thompson, K. (1995). *The electronic comparative plant ecology.* London: Chapman & Hall.

Janzen, D. H. (1970). Herbivores and the number of tree species in tropical forests. *American Naturalist* **104**, 501–528.

Kelly, D. (1982). Demography, population control and stability in short-lived plants of chalk grassland. Unpublished Ph.D. thesis, University of Cambridge.

Law, R. & Watkinson, A. R. (1989). Competition. In *Ecological concepts* (ed. J. M. Cherrett), pp. 243–285. Oxford: Blackwell Scientific Publications.

Leishman, M. R. & Westoby, M. (1994). Hypotheses on seed size: tests using the semiarid flora of western New South Wales, Australia. *American Naturalist* **143**, 890–906.

Leishman, M. R., Westoby, M. & Jurado, E. (1995). Correlates of seed size variation: a comparison of five temperate floras. *Journal of Ecology* **83**, 517–530.

Lloyd, D. G. (1987). Selection of offspring size at independence and other size-versus-number strategies. *American Naturalist* **129**, 800–817.

Lokesha R., Hedge S. G., Uma Shaanker R. & Ganeshaiah J. N. (1992). Dispersal mode as a selective force in shaping the chemical composition of seeds. *American Naturalist* **140**, 520–525.

Mazer, S. J. (1989). Ecological, taxonomic, and life-history correlates of seed mass among Indiana dune angiosperms. *Ecological Monographs* **59**, 153–175.

Parker, G. A. & Begon M. (1986). Optimal egg size and clutch size: effects of environment and maternal phenotype. *American Naturalist* **128**, 573–592.

Ponquett, R. T., Smith, M. T. & Ross, G. (1992). Lipid autoxidation and seed ageing: putative relationships between seed longevity and lipid stability. *Seed Science Research* **2**, 51–54.

Purvis, A., & Rambaut A. (1995). Comparative analysis using independent contrasts (CAIC): an Apple Macintosh application for analysing comparative data. *Computer Applications in the Biosciences* **11**, 247–251.

Rabinowitz, D. & Rapp, J. K. (1981). Dispersal abilities of seven sparse and common grasses from a Missouri prairie. *American Journal of Botany* **68**, 616–624.

Rees, M. (1993). Trade-offs among dispersal strategies in the British flora. *Nature* **366**, 150–152.

Rees, M. (1994). Delayed germination of seeds: a look at the effects of adult longevity, the timing of reproduction, and population age/stage structure. *American Naturalist* **144**, 43–64.

Rees, M. (1995). EC-PC comparative analyses? *Journal of Ecology* **83**, 891–892.

Rees, M. & Long, M. J. (1992). Germination biology and the ecology of annual plants. *American Naturalist* **139**, 484–508.

Rees, M., Grubb, P. J. & Kelly, D. (1996). Quantifying the effects of competition and spatial heterogeneity on the structure and dynamics of a four species guild of dune annuals. *American Naturalist* **147**, 1–32.

Rees, M. & Long, M. J. (1993). The analysis and interpretation of seedling recruitment curves. *American Naturalist* **141**, 233–262.

Rees, M. & Westoby, M. (1997). Game-theoretical evolution of seed mass in multi-species ecological models. *Oikos* **78**, 116–126.

Roberts, H. A. (1964). Emergence and longevity in cultivated soil of seeds of some annual weeds. *Weed Research* **4**, 296–307.

Roberts, H. A. (1979). Periodicity of seedling emergence and seed survival in some Umbelliferae. *Journal of Applied Ecology* **16**, 195–201.

Roberts, H. A. & Boddrell, J. E. (1983). Seed survival and periodicity of seedling emergence in eight species of Cruciferae. *Annals of Applied Biology* **103**, 301–309.

Salisbury, E. J. (1942). *The reproductive capacity of plants*. London: G. Bell and Sons.

Salisbury, E. J. (1974). Seed size and mass in relation to environment. *Proceedings of the Royal Society* **186**, 83–88.

Silvertown, S. W. (1981). Seed size, life span, and germination date as coadapted features of plant life history. *American Naturalist* **118**, 860–864.

Smith, C. C. & Fretwell, S. D. (1974). The optimal balance between size and number of offspring. *American Naturalist* **108**, 499–506.

Stace, C. (1991). *New flora of the British Isles*. Cambridge University Press.

Thompson, K. (1987). Seeds and seed banks. *New Phytologist* **106**, 23–34.

Thompson, K. & Rabinowitz, D. (1989). Do big plants have big seeds? *American Naturalist* **133**, 722–728.

Tilman, D. (1988). *Plant strategies and the dynamics and structure of plant communities*. Monographs in Population Biology 26. Princeton University Press.

Tilman, D. & Pacala, S. W. (1993). The maintenance of species richness in plant communities. In *Species diversity in ecological communities: historical and geographical perspectives* (ed. R. E. Ricklefs & D. Schluter), pp. 13–25. University of Chicago Press.

Venable, D. L. & Brown, J. S. (1988). The selective interactions of dispersal, dormancy, and seed size as adaptations for reducing risk in variable environments. *American Naturalist* **131**, 360–384.

Venable, D. L., Pake, C. E. & Caprio, A. C. (1993). Diversity and coexistence of Sonoran desert winter annuals. *Plant Species Biology* **8**, 207–216.

Vincent, T. L., Cohen, Y. & Brown, J. S. (1993). Evolution via strategy dynamics. *Theoretical Population Biology* **44**, 149–176.

Watkinson, A. R. & Davy, A. J. (1985). Population biology of salt marsh and sand dune annuals. *Vegetatio* **62**, 487–497.

Westoby, M., Jurado, E. & Leishman, M. (1992). Comparative evolutionary ecology of seed size. *TREE* **7**, 368–372.

Wilson, A. M. & Thompson, K. (1989). A comparative study of reproductive allocation in 40 British grasses. *Functional Ecology* **3**, 297–302.

8 • Comparative ecology of seed size and dispersal

Mark Westoby, Michelle Leishman and Janice Lord

1 Introduction

Between about 1965 and 1975, ecology changed research style. People were becoming more acutely aware that patterns as they are found in nature – 'natural experiments' – are hard to interpret because factors are confounded. Even though every effort is made after the fact to account for possible cross-correlations, a natural experiment is not capable of proving which of two confounded influences is a direct cause and which a secondary correlate. In this situation the idea that competitors could be removed, or predators excluded with cages, in a properly replicated and randomized manner in natural environments, caught the imagination of a generation. Manipulative experiments have remained the research style that defined leadership in ecology for the past 25 years up to the present (Roush 1995).

Now during the 1990s, research styles are in transition again. Many hundreds of experiments have accumulated on competition and other inter-actions. Often they have different outcomes, depending on the species and the situations involved. The problem of scaling-up from experimental results has become acute. Therefore over the next 10–20 years putting species into comparative context will be the key to research ecology. Comparative ecology will be crucial for improved meta-analysis of the large numbers of experiments that have already accumulated, for generalizing from species and situations that have been the subject of experiments, and for intelligent selection of species for further experiments. This chapter concentrates on seed mass. Our underlying interest, though, is in understanding the whole attribute constellation, with a view to plant ecological strategy schemes.

Seed mass varies greatly between plant species (Harper *et al.* 1970; Westoby *et al.* 1992), against a background of comparatively narrow variation within species. Standard deviations for log seed mass between species within a vegetation type are typically about 1.0, in other words, ± one SD spans a

100-fold range (unpublished data, seven floras ranging from 0.76 to 1.14). In contrast within-species SDs have medians about 0.3, in other words ± one SD spans about a four-fold range (Michaels *et al.* 1988). Moreover, most of the variation within species occurs among different seeds on the same mother plant (Michaels *et al.* 1988, Obeso 1993), indicating it is largely due to vagaries in the developmental process, rather than being heritable.

Ecological research to understand the variation in seed mass (or any other attribute) between species includes four distinct though complementary questions:

1. In what patterns is seed mass correlated with other components of the attribute constellation of plant species?
2. To what extent can larger seed mass be shown experimentally to support better seedling establishment in the face of various hazards?
3. In what way is seed mass correlated with the environmental conditions under which a species' seedlings naturally establish?
4. To what extent does seed mass variation take the form of relatively recent evolutionary divergences, between species within genera, versus the form of older divergences, between say orders or families?

This chapter deals with the first three questions in sequence, commenting where possible on the relationship to phylogeny. Seed mass differences are often conservative between genera or families (Hodgson & Mackey 1986; Mazer 1990; Peat & Fitter 1994; Lord *et al.* 1995).

Our approach through four questions is different, and deliberately so, from that of a number of other contributors to this volume (e.g. Crawley *et al.*, Franco & Silvertown, Silvertown & Dodd, all this volume). These authors advocate a single methodology, which they regard as 'correct'. A phylogenetic tree, plus the information on extant species at the tree tips, is used to reconstruct trait changes along branches of the tree. One then tests for correlated evolutionary change – trait A changing in a particular direction along the same branch where trait B changes, more often than expected by chance. This procedure estimates the number of separate divergences for a trait combination. The question addressed is a variant of question 4 above. Our disagreement with these authors is not about the validity of the correlated-change test, but about interpreting its outcome. They regard the test for correlated change as producing the correct answer, whereas tip correlations are flawed because related species may share traits due to 'common ancestry, not adaptation', leading to 'pseudoreplication' (Silvertown & Dodd, this volume). Regrettably, at this meeting they have continued to represent correlated-change tests as the only path to correct answers, and have not qualified this by warning readers that their claim is controversial (see the 1995 forum in *Journal of Ecology*: Ackerly & Donoghue 1995; Fitter 1995; Harvey *et al.* 1995, *b*; Rees 1995; Westoby *et al.* 1995*a*, *b*, *c*).

In fact, it is meaningless to assert the correctness or otherwise of any statistical test except in relation to a particular question or hypothetical model of the process underlying a pattern (Harvey & Pagel 1991). The generic claim to correctness is actually a claim that only one question or model is worth investigating; similarly with assertions that related species are not independent samples. One cannot say this in the abstract: one might as well say that species are not independent because both have wings on their seeds, or have any other thing in common. Assertions about independence are actually assertions about the model that is being tested.

The test adopted by Silvertown and others estimates degrees of freedom correctly as a test of correlated evolutionary change along the same tree branch. But the correlated-change test is simply incorrect for several other models that biologists might be interested in. It is incorrect for the model that change in one trait is a prerequisite for change in another (e.g. Barrett *et al.*, Donoghue & Ackerly, this volume). Tests for this model need to look for changes in A along tree branches that precede changes in B, not along the same branch. The correlated-change test is also incorrect for models that regard stasis or maintenance of a trait as meaningful events, just as meaningful as change. Trait maintenance models should be investigated when traits are suspected to be responsive to selection, or when 'traits' are actually ecological outcomes, such as geographic range (Kelly & Woodward 1996) or distribution on one soil type rather than another (van Groenendael *et al.*, this volume). Ecological outcome 'traits' (see question 3 above) arguably have to be maintained in each generation by diaspores establishing in some habitats but failing in others. For these models, the correlated-change test incorrectly underestimates the degrees of freedom, by ruling that maintenance of a trait is not evidence for anything, whereas change in a trait is evidence. The correlated-change test is also irrelevant to models about a trait's ecological function in the present day. For the extreme case of a trait that has originated only once, a correlated-change test will find only a single divergence, and therefore will always accept a null hypothesis that nothing meaningful has happened. Plainly it is wrong to conclude that a trait that has originated only once cannot have a meaningful ecological function, so this example illustrates the futility of using a correlated-change test to investigate present-day functionality.

In summary, there are a number of legitimate questions that can be asked (see the four above, also the slightly different list in Westoby *et al.* 1995*b*). Accordingly no single statistical procedure is 'correct' for all purposes, and conversely procedures should not be stigmatized as 'incorrect' without paying careful attention to how the interpretation is phrased. It is interpretations, rather than statistical procedures, that are correct or incorrect.

Perhaps most fundamentally, comparative datasets provide only correlative evidence, and what is more, correlative evidence in which different causes are confounded (Westoby *et al.* 1995*c*). While the correlative evidence deserves investigating for its consistency with a given model of causation, in the final analysis we cannot determine causation from cross-correlated patterns, no matter how sophisticated the procedures for correcting one variable for another. Confidence that we have understood causation correctly can only come from a mixture of types of evidence, including experiments on ecological outcomes, and understanding of physiological and developmental mechanisms, as well as comparative data. This paper brings together different types of evidence about the ecology and evolution of seed mass, and assesses the extent to which these different types of evidence present a consistent picture.

2 Seed mass in relation to other attributes of species

2.1 Dispersal biology

The relationship between seed mass and dispersal mode is broadly similar in five temperate floras spanning three continents (Figure 8.1*a*). The nature of the relationship is that seeds above 100 mg tend to be adapted for dispersal by vertebrates, seeds below 0.1 mg tend to be unassisted, but between 0.1 and 100 mg all dispersal modes are feasible (Hughes *et al.* 1994). Correspondingly, the relationship has substantial r^2 (0.29), but at the same time fully 71% of variation in log seed mass occurs within dispersal modes. While the differences between floras in the shape of the relationship are significant, they are about ten times smaller than the consistent element of the pattern (flora × dispersal mode interaction $r^2 = 0.03$ *vs* dispersal mode main effect $r^2 = 0.29$, Leishman *et al.* 1995).

Most but not all of the difference in average seed mass between dispersal modes is associated with shifting family representation (85% averaged across six floras; Lord *et al.* 1995). Generally, seed mass is quite phylogenetically conservative, with 55% of log seed mass variation between orders or above,

12% between families within orders, 26% between genera within families and 8% between species within genera (averages across six floras; Lord *et al.* 1995).

2.2 Plant height and growth form

As was the case for dispersal mode, the relationship of seed mass to growth form is reasonably consistent between different floras (Fig 1*b*; growth form main effect $r^2 = 0.20$ *vs* growth form × flora interaction $r^2 = 0.02$). Climbers and woody plants have average seed mass about one order of magnitude larger than forbs and graminoids (Figure 8.1*b*).

Growth form and dispersal mode are in turn correlated, so there is some overlap between the proportion of seed mass correlated with growth form and the proportion correlated with dispersal mode, but each is correlated with a substantial portion independently of the other. Taken together, growth form and dispersal mode were capable of predicting between 21 and 47% of log seed mass variation in five different floras (Leishman *et al.* 1995).

Most but not all of the difference in average seed mass between growth forms is associated with shifting family representation (93% averaged across six floras; Lord *et al.* 1995).

2.3 Specific leaf area

For unclear reasons, species with large seed mass tend to have lower specific leaf area (SLA), leaf area per leaf dry mass (Figure 8.2). Species vary along a spectrum from long-lived, evergreen leaves with low SLA to short-lived leaves with large SLA. Species at the low-SLA end of the spectrum typically have lower N contents per unit mass (though not necessarily per unit area), may allocate substantial mass to tannins, phenols or other defensive compounds, and can achieve only slow relative growth rate (RGR) even under favourable conditions. Species with high SLA are deploying more light-catching area per unit photosynthate invested, have faster potential RGR (Garnier 1992; Lambers & Poorter 1992; Reich *et al.* 1992, Saverimuttu & Westoby 1996*a*), and the resulting faster turnover of plant parts permits a more flexible response to the spatial patchiness of light and soil resources (Grime 1994).

Low SLA and the associated nexus of attributes has been widely identified as a syndrome of adaptation to unfavourable sites (e.g. Grime 1977; Leps *et al.* 1982; Loehle 1988; Reich *et al.* 1992; Aerts & van der Peijl 1993; Chapin *et al.*

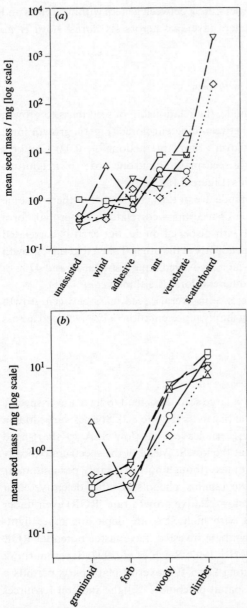

Figure 8.1. Relationship of seed mass to (a) dispersal mode and (b) growth form in five temperate floras, (open circles: western New South Wales; open squares: central Australia; closed triangles: Sydney; open triangles: Indiana Dunes; open diamonds: Sheffield), after Leishman *et al.* (1995).

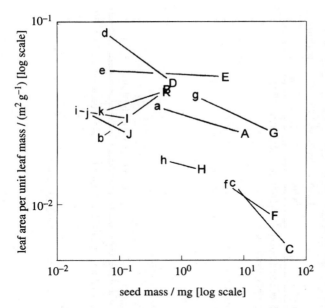

Figure 8.2. Relationship of SLA measured during seedling growth to seed mass, 11 PICs (data from Saverimuttu & Westoby, (1996*b*)).

1993; van der Werf *et al.* 1993). Historically, attention first focussed on potential RGR of a species (Grime & Hunt 1975), measured on exponentially growing seedlings given plentiful water and nutrients, as the expression of this syndrome. A spectrum of potRGR is central to theories about comparative plant ecology and vegetation dynamics (Grime 1977; Tilman 1990; Keddy & McLellan 1992). It has now become apparent that a trade-off between maximizing light-capturing area and maximizing leaf durability underlies this spectrum. SLA accounts for most potRGR variation between species, though LWR (leaf as proportion of plant mass) sometimes plays a role, especially in comparisons between growth forms.

Species-pairs in Figure 8.2 were chosen to form phylogenetically independent contrasts (PICs). Pairs were contrasted for seed mass, and each pair represented a phylogenetic divergence independent of the other pairs. The SLA–seed mass relationship is present both as a tip correlation considered across all the species, and as a tendency of lines connecting PIC pairs to angle downwards. PICs were mostly within genera or families, so the relationship resides both in differences between superorders, orders and families, and also in differences within genera and families. Some of the effect is associated with

Table 8.1. *Experiments that manipulated the environment in order to test the proposition that seedlings from species having larger seeds perform better than those from species having smaller seeds, under various hazards*

Hazard	Larger-seeded species performed better	Larger-seeded species did not perform better
Competition from established vegetation	Gross & Werner 1982; Gross 1984; McConnaughay & Bazzaz 1987; Bakker 1989; Reader 1993; Burke & Grime 1996	Thompson & Baster 1992
Deep shade	Grime & Jeffrey 1965; Leishman & Westoby 1994a; Saverimuttu & Westoby 1996b; Osunkoya et al. 1994	Augspurger 1984[a]; Saverimuttu & Westoby 1996[b]
Defoliation	Armstrong & Westoby 1993	
Mineral nutrient shortage	Lee & Fenner 1989; Jurado & Westoby 1992	
Depth under soil or litter	Gulmon 1992; Peterson & Facelli 1992; Vasquez-Yanes & Orozco-Segovia 1992	
Dry environments	Leishman & Westoby 1994b (glasshouse)	Leishman & Westoby 1994b (field)[c]

[a] PAR at $17.5 \, \mu\text{mol m}^{-2} \, \text{s}^{-1}$ was above compensation point. Mortality occurred not at cotyledon stage but mainly between weeks 5–30, was mainly due to disease, and was correlated with lower wood density.
[b] Seedlings grown in full light to first-leaf stage, then transferred to dense shade.
[c] Very severe temperature conditions, low rates of emergence and survival even in the best-watered treatment.

species of larger seed mass belonging to taller or more perennial growth forms, but several of the PICs are within growth forms.

Seed mass had previously been found correlated with slower potRGR in most datasets (summaries in Shipley & Peters 1990, Westoby *et al.* 1992; see also for use of PICs Marañón & Grubb 1993; Swanborough & Westoby 1996), so the relationship to SLA was predictable, though the basis for it is not yet understood.

3 Seedling outcomes associated with larger seed size

In experiments where particular environmental hazards are deliberately varied independently of other factors, seedlings from larger-seeded species have often been shown to perform better than those from smaller-seeded species (Table 8.1). These experiments are directed at demonstrating present day functionality. Taken together, they show that larger seed size can be functional in the present day in relation to a wide range of hazards.

Two studies have used species selected to form PICs. Capacity to survive 95% excision of cotyledons was associated with seed mass within genera and families, but between families and orders there was little relationship to seed mass, probably because differences in seedling morphology became more important (Armstrong & Westoby 1993). Cotyledon-stage longevity under dense shade (Saverimuttu & Westoby 1996*b*) was associated with seed mass at all levels, both in PICs within genera and families, and also in older evolutionary divergences across PICs between families and orders.

The exceptions where larger-seeded species did not perform better are of interest. For drought, the exception occurred under very hot field conditions, when survival to emergence was low for all seed sizes. For competition with established vegetation, the exception did not span a very wide range of seed size. For shading, the exceptions occurred after the initial phase of deployment from seed reserves into seedling. This evidence is relevant to mechanisms, which are discussed under the next heading.

4 A common mechanism for tolerating different hazards?

Might the better performance of larger-seeded species, under a range of different hazards (Table 8.1), be mediated through common machinery? There would seem to be three candidates for machinery that might be operating:

1. *Seedling size effect*: because larger seeds give rise to larger seedlings immediately after germination, they may reach deeper into the soil to better water supplies, or higher into the air to a better photosynthetically active radiation (PAR) level.
2. *Reserve effect*: extra metabolic resources in larger seeds may serve to support carbon deficits.
3. *Metabolic effect*: since larger-seeded species tend to have slower potential RGR (for unknown reasons), perhaps they have slower respiration rates or otherwise consume metabolic resources more slowly, and it is this rather than seed size that gives them longer survivorship under various hazards.

We have positive evidence that the seedling size effect and the metabolic effect cannot be a common machinery, though they might still be relevant in particular situations. The seedling size effect is not capable of accounting for the improved survivorship of seedlings where gradients of resource supply away from the soil surface are not relevant. It cannot account for the longer survival when deprived of any access to mineral nutrients or shaded experimentally below the compensation point, nor for the outcome of the defoliation experiment.

Saverimuttu & Westoby (1996*b*) found evidence against a metabolic rate mechanism. Among seedlings placed in deep shade below the compensation point immediately following germination, at cotyledon stage, larger-seeded species tended to survive longer. But slow dark respiration rates and slow potential RGR in full light were not such good predictors of shade longevity as was seed reserve mass itself. Further, when seedlings were grown in full light through to first-leaf stage (a stage when they had fully embarked on exponential growth and seed reserves had been fully deployed), and only then were transferred to deep shade, longevity in deep shade was not well correlated with seed mass. These seedlings died faster than cotyledon-stage seedlings, the difference being apparent especially for species with larger seed mass. These results indicate that the advantage of larger seeds applies only while the seed reserves are being deployed into the seedling, and does not persist into later seedling life. Augspurger's (1984) finding no relationship between seed mass and survival was also consistent with this conclusion. PAR at 17.5 μmol m^{-2} s^{-1} was probably above compensation point, and seedling mortality occurred mainly between weeks 5–30 rather than in earlier weeks, was mainly due to disease and was correlated with lower wood density.

Three points may be made in summary of present knowledge about the mechanism or mechanisms by which larger seed size permits better seedling performance in the face of different hazards.

First, if there is a single underlying mechanism, it must be the reserve effect, as there is clear evidence against a size effect or a metabolic effect, at least for some types of hazard. But why should larger seeds have more resources available to them *relative to the size of the seedling to be supported*? The reserve effect would seem to require that in larger seeded species, a greater proportion of the seed's stored resources are in some sense uncommitted during deployment, and capable of being used to support respiration under carbon deficit. However, this proposition, that during deployment from seed into seedling, larger-seeded species hold a greater proportion of seed reserves in forms where they can be retrieved to support respiration, has yet to be tested directly.

Second, if the reserve effect is a single unifying mechanism, larger seed size might initially arise due to selection by one type of hazard, but would subsequently be ecologically functional in relation to another type – indeed, in relation to *all* other types of hazard. This illustrates the general principle that forces favouring the initial emergence of a trait are not necessarily the same as those maintaining it. Conversely, experiments about present-day functionality are not eligible evidence for researchers seeking to infer the origins of traits, unless they are willing to make an argument that past ecological circumstances have much in common with those of the present day.

Third, if the mechanism is via a reserve effect, the benefits conveyed by large-seededness are temporary. At some point between cotyledon stage and true-leaf stage, reserves are irreversibly deployed into the body of the seedling and large-seeded species have no further advantage. An important implication is that large-seededness would only be expected to convey benefits in relation to hazards that are temporary, where there is at least some probability that conditions will improve after a while.

5 Patterns of seed mass in relation to environmental factors

The only clearly established pattern in the field is that species maintaining populations where establishment in the shade is required tend to have larger seeds (Table 8.2). Any relationship to drought risk is much more marginal. Mazer (1989) did not find it. Baker (1972) is often cited as having found it, but his California data actually show little trend across moisture classes 1–5, with

Table 8.2. *Publications reporting whether species whose seeds establish under particular environmental conditions tend to have larger seed mass*

	Yes	Uncertain	No
Shaded	Salisbury 1942; Foster & Janson 1985; Mazer 1989; Hammond & Brown 1995	Metcalfe & Grubb 1995	Westoby *et al.* 1990; Hammond & Brown 1995
Low soil nutrients			
Droughted		Salisbury 1942; Baker 1972	Mazer 1989

any tendency to smaller seeds present only in intermittently or permanently flooded sites of moisture class 6 (Westoby et al. 1992).

The absence of any pattern in relation to soil nutrients is consistent with the idea that advantages ought to apply only where the hazard may be temporary. Similarly, it seems reasonable to regard shading as a hazard that is potentially temporary. Although an increase in longevity below the compensation point from (say) 15 to 30 days only provides a small absolute probability that a tree will fall and a light gap open up while the seedling is still alive, the probability is nevertheless doubled. With regard to drought, dry soil is plainly capable of being a temporary rather than a continuing problem for a seedling. It is not clear, though, whether environments with lower annual rainfall pose a greater hazard of drought during establishment. Seedlings will germinate only at particular times of year, sometimes only after soil-saturating initial rainfall. Possibly in the period of 1–2 weeks between germination and independence from seed reserves, seedlings in high rainfall zones are just as much at risk from drought as seedlings in arid zones.

6 State of evidence on benefits of larger seed mass in relation to different hazards

In summary, experiments show that larger seed mass can be functional in relation to a wide range of environmental hazards, but only in relation to establishment in the shade does there appear to be any strong or consistent distribution pattern in the field. (Patterns in the field have, however, hardly been investigated in relation to seedling physical damage risk and soil nutrients, and the evidence remains indecisive in regard to drought risk.) Even in relation to shading, the mechanism is not understood in depth. For most benefits, the benefit and the larger seed mass are modulated together to some extent within genera and families, but in addition much of the correlation is underpinned by older evolutionary divergences, between families, orders or superorders.

7 The wide spread of coexisting seed sizes implies they are determined game-theoretically rather than by simple optimization in relation to environment

Seed mass is thought to be shaped as a size-number compromise (Harper et al. 1970; Harper 1977; Willson 1983; Westoby et al. 1992). This idea is made graphical in the Smith & Fretwell (1975) model. The Smith–Fretwell function describes a diminishing returns relationship between seed reserve mass and a seedling's chances of establishing and eventually reproducing. The prediction follows that there should be a single best (fastest λ) seed size. If species have different seed sizes (as indeed they do), this should express the fact that they are evolving under different Smith–Fretwell functions.

Since the 1970s, most research on seed mass has been directed at understanding how larger versus smaller seeds perform under various hazards. The implied agenda has been to look for differences in the Smith–Fretwell functions. This implied agenda now needs to be reconsidered. It has become apparent that there is a wide range of seed mass strategies occurring interspersed within most vegetation types, compared with surprisingly small differences in the mean between environments. Across five datasets, variation within accounted for 96% of total variance, differences between for only 4% (Leishman et al. 1995). Given that the datasets ranged from arid woodlands, through coastal rainforest and sclerophyll rainforest, to cool temperate closed grasslands, it is hard to reconcile this with the idea that the prevalence of different physical circumstances during seedling establishment is the main force favouring one seed size versus another.

Recently, game-theoretic methods have been applied to seed mass (Geritz 1995; Rees & Westoby 1997), superseding the tacit assumption that the competitive context can be subsumed into the shape of the Smith–Fretwell function. As we should have expected, the game-theoretic approach makes a fundamental difference. Suppose we assume that seeds fall at random into patches just big enough to accommodate a single adult plant. Competition determines that within a patch, a larger seed will be successful at the expense of smaller seeds. Then, no single seed mass strategy constitutes an evolutionary stable strategy (ESS). A strategy set consisting of a single, medium seed size, as might be predicted from Smith–Fretwell, can be invaded by larger seeds because they win competition in those patches where they occur. Smaller seeds can invade, because they are produced in greater numbers and there will be some patches that are reached by them but not by any larger seeds. The upper bound to the strategy set is where the seeds produced are so few that $\lambda \leqslant 1$. The lower bound is at the seed mass that maximizes λ after density-independent mortality during dispersal and germination, but before competition. A broad mix of seed-size strategies is expected between those bounds. This result echoes conclusions from models formulated as lotteries (Ågren & Fagerstrom 1984) and as a colonization-competition trade-off (Tilman 1994), though those models were not explicitly about seed mass.

So a game-theoretic approach is capable of predicting a wide mix of seed-mass strategies coexisting, and this actually occurs for reasons otherwise unexplained. But the game-theoretic models only apply literally where species composition is determined to a substantial extent by competition between seedlings. This might be plausible for vegetation of annuals, fire-prone vegetation, arid-zone vegetation where seedling establishment occurs in bursts after major rain, and forest vegetation with gap dynamics, but most plant ecologists would not yet be ready to believe that it applies universally in all vegetation types.

8 Outstanding matters that are not yet clear

The present state of knowledge includes both some matters that seem reasonably settled and consistent, and other matters that are by no means clear. The most important unclear questions are:

1. What is the mechanism by which larger seeds support seedlings better under different hazards? Is it the case that larger seeds make a less complete commitment of their reserves during seedling deployment, holding more available to support the seedling during temporary carbon deficits? Does the mechanism imply that the benefit of larger seed mass will be generic, applying under a wide range of different possible hazards? And, might it be possible to find an indicator of this incomplete commitment that would serve as a better predictor than seed mass itself?
2. What is the reason for the correlation of larger seed mass with lower SLA and its associated nexus of attributes related to vegetative growth? Some of the raw correlation is associated with taller growth forms, but the correlation is also present within growth forms.
3. What might account for the very broad spread of seed mass within vegetation types, compared with the minor differences in mean seed mass between very different vegetation types? Four hypotheses seem possible candidates: (a) species germinate under different circumstances, and therefore face a sufficiently broad variety of establishment hazards within each vegetation type – a sufficient diversity of Smith–Fretwell functions – to account for the spread of seed mass; (b) many species in any vegetation type might occur as sink populations with internal $\lambda < 1$, but supported by dispersal from source populations in other habitats; (c) Game-theoretic mixture of strategies for seed mass as in the models of Geritz (1995) and Rees & Westoby (1997) – this implies that species composition is substantially determined by competition among seedlings; (d) game-theoretic

mixture of strategies for some other attribute, with seed mass secondarily correlated with this other attribute.

9 Some matters that seem sufficiently clear

As well as pointing out matters that remain unresolved and where further research is needed, it bears emphasizing that we have also substantial knowledge about seed mass, reasonably firmly established:

1. Larger seed mass does convey benefits in seedling establishment, under a wide variety of circumstances.
2. Seed mass is correlated with several other plant attributes, those of greatest importance in defining the ecology of a species being height or growth form, dispersal mode and SLA or potential RGR. These relationships have substantial r^2, and their patterns appear consistent in floras from very different environments and having different phylogenetic backgrounds. Attributes of the regenerative phase of life histories are not in general well correlated with attributes of vegetative growth (Grime *et al.* 1988; Shipley *et al.* 1989; Leishman & Westoby 1992). But seed mass is connected both to dispersal biology and to growth form and SLA, and occupies a pivotal position in the constellation of attributes that determines under which environmental opportunities a species is most competitive.

We thank members of the Macquarie Ecology Discussion Group, David Ackerly, Spencer Barrett, Toby Fagerstrom and Lawrence Harder for valuable discussion. We are grateful for funding from the Australian Research Council and the New Zealand Foundation for Research Science and Technology. This is contribution no. 214 from the Research Unit for Biodiversity and Bioresources.

References

Ackerly, D. D. & Donoghue, M. J. (1995). Phylogeny and ecology reconsidered. *Journal of Ecology*, **83**, 730–732.

Aerts, R. & van der Peijl, M. J. (1993). A simple model to explain the dominance of low productive perennials in nutrient-poor habitats. *Oikos* **66**, 144–147.

Ågren, G. I. & Fagerstrom, T. (1984). Limiting dissimilarity in plants: randomness prevents exclusion of species with similar competitive abilities. *Oikos* **43**, 369–375.

Armstrong, D. P. & Westoby, M. (1993). Seedlings from large seeds tolerate defoliation better: a test using phylogenetically independent contrasts. *Ecology* **74**, 1092–1100.

Augspurger, C. K. (1984). Light requirements of neotropical tree seedlings: a comparative study of growth and survival. *Journal of Ecology* **72**, 777–795.

Baker, H. G. (1972). Seed weight in relation to environmental conditions in California. *Ecology* **53**, 997–1010.

Bakker, J. P. (1989). *Nature management by grazing and cutting.* Dordrecht: Kluwer.

Burke, M. J. W. & Grime, J. P. (1996). An experimental study of plant community invasibility. *Ecology,* **77**, 776–790.

Chapin, F. S. III, Autumn, K. & Pugnaire, F. (1993). Evolution of suites of traits in relation to environmental stress. *American Naturalist* **142**, S78–S92.

Fitter, A. H. (1995). Interpreting quantitative and qualitative characteristics in comparative analyses. *Journal of Ecology* **83**, 730.

Foster, S. A. & Janson, C. H. (1985). The relationship between seed size and establishment conditions in tropical woody plants. *Ecology,* **66**, 773–780.

Garnier, E. (1992). Growth analysis of congeneric annual and perennial grass species. *Journal of Ecology* **80**, 665–675.

Geritz, S. A. H. (1995). Evolutionarily stable seed polymorphism and small-scale spatial variation in seedling density. *American Naturalist* **146**, 685–707.

Grime, J. P. (1977). Evidence for the existence of three primary strategies in plants and its relevance to ecological and evolutionary theory. *American Naturalist* **111**, 1169–1194.

Grime, J. P. (1994). The role of plasticity in exploiting environmental heterogeneity. In *Exploitation of environmental heterogeneity by plants: ecophysiological processes above- and below-ground* (ed. M. M. Caldwell & R. W. Pearcy), pp. 1–19. New York: Academic Press.

Grime, J. P. & Hunt, R (1975). Relative growth rate: its range and adaptive significance in a local flora. *Journal of Ecology* **63**, 393–422.

Grime, J. P. & Jeffrey, D. W. (1965). Seedling establishment in vertical gradients of sunlight. *Journal of Ecology* **53**, 621–634.

Grime, J. P., Hodgson, J. G. & Hunt, R. (1988). *Comparative plant ecology: a functional approach to common British species.* London: Unwin-Hyman.

Gross K. L. (1984). Effects of seed size and growth form on seedling establishment of six monocarpic perennial plants. *Journal of Ecology* **72**, 369–387.

Gross K. L. & Werner P. A. (1982). Colonizing abilities of 'biennial' plant species in relation to ground cover: implications for their distributions in a successional sere. *Ecology* **63**, 921–931.

Gulmon S. L. (1992). Patterns of seed germination in Californian serpentine grassland. *Oecologia* **89**, 27–31.

Hammond, D. S. & Brown, V. K. (1995). Seed size of woody plants in relation to disturbance, dispersal, soil type in wet neotropical forests. *Ecology* **76**, 2544–2561.

Harper J. L. (1977). *Population biology of plants.* New York: Academic Press.

Harper J. L., Lovell P. H. & Moore K. G. (1970). The shapes and sizes of seeds. *Annual Review of Ecology and Systematics* **1**, 327–356.

Harvey, P. H. & Pagel, M. D. (1991). *The comparative method in evolutionary biology*. Oxford University Press.

Harvey, P. H., Read, A.F. & Nee, S. (1995a) Why ecologists need to be phylogenetically challenged. *Journal of Ecology*, **83**, 535–536.

Harvey, P.H, Read, A.F. & Nee, S. (1995b). Further remarks on the role of phylogeny in comparative ecology. *Journal of Ecology*, **83**, 735–736.

Hodgson, J. G. & Mackey, J. M. L. (1986). The ecological specialization of dicotyledonous families within a local flora: some factors constraining optimization of seed size and their possible evolutionary significance. *New Phytologist* **104**, 497–515.

Hughes, L., Dunlop, M., French, K. *et al* (1994). Predicting dispersal spectra: a minimal set of hypotheses based on plant attributes. *Journal of Ecology* **82**, 933–950.

Jurado, E. & Westoby, M. (1992). Seedling growth in relation to seed size among species of arid Australia. *Journal of Ecology* **80**, 407–416.

Keddy, P. A. & McLellan, P. (1992). Centrifugal organization in forests. *Oikos* **59**, 75–84.

Lambers, H. & Poorter, H. (1992). Inherent variation in growth rates between higher plants: a search for physiological causes and higher consequences. *Advances in Ecological Research* **23**, 188–261.

Lee, W. G. & Fenner, M. (1989). Mineral nutrition allocation in seeds and shoots of 12 *Chionochloa* sp. in relation to soil fertility. *Journal of Ecology* **77**, 704–716.

Leishman, M. R. & Westoby, M. (1992). Classifying plants into groups on the basis of associations of individual traits – evidence from Australian semi-arid woodlands. *Journal of Ecology* **80**, 417–424.

Leishman, M. R. & Westoby, M. (1994a). The role of large seed size in shaded conditions: experimental evidence. *Functional Ecology* **8**, 205–214.

Leishman, M. R. & Westoby, M. (1994b). The role of large seeds in seedling establishment in dry soil conditions – experimental evidence from semi-arid species. *Journal of Ecology* **82**, 249–258.

Leishman, M. R., Westoby, M. & Jurado, E. (1995). Correlates of seed size variation: a comparison among five temperate floras. *Journal of Ecology* **83**, 517–530.

Leps, J., Osborna-Kosinova, J. & Rejmanek, K. (1982). Community stability, complexity and species life-history strategies. *Vegetatio* **50**, 53–63.

Loehle, C. F. (1988). Tree life history strategies: the role of defenses. *Canadian Journal of Foraminiferal Research* **18**, 209–222.

Lord, J., Westoby, M. & Leishman, M. R. (1995). Seed size and phylogeny in six temperate floras: constraints, niche conservatism and adaptation. *American Naturalist* **146**, 349–364.

Marañón, T., & Grubb, P. J. (1993). Physiological basis and ecological significance of the seed size and relative growth rate relationships in Mediterranean annuals. *Functional Ecology* **7**, 591–599.

Mazer, S. J. (1989). Ecological, taxonomic and life history correlates of seed mass among Indiana dune angiosperms. *Ecological Monographs* **59**, 153–175.

Mazer, S. J. (1990). Seed mass of Indiana Dune genera and families: taxonomic and ecological correlates. *Evolutionary Ecology* **4**, 326–357.

McConnaughay, K. D. M. & Bazzaz, F. A. (1987). The relationship between gap size and performance of several colonizing annuals. *Ecology* **68**, 411–416.

Metcalfe, D. J. & Grubb, P. J. (1995). Seed mass and light requirement for regeneration in South-East Asian rain forest. *Canadian Journal of Botany* **73**, 817–826.

Michaels H .J., Benner, B., Hartgerink, A. P. *et al.* (1988). Seed size variation: magnitude, distribution and ecological correlates. *Evolutionary Ecology* **2**, 157–166.

Obeso, J. R. (1993). Seed mass variation in the perennial herb *Asphodelus albus*: sources of variation and position effect. *Oecologia* **93**, 571–575.

Osunkoya, O. O., Ash, J. E., Hopkins, M. S. & Graham, A. W. (1994). Influence of seed size and seedling ecological attributes on shade-tolerance of rainforest tree species in northern Queensland. *Journal of Ecology* **82**, 149–163.

Peat, H. J. & Fitter, A. H. (1994). Comparative analyses of ecological characteristics of British angiosperms. *Biological Reviews of the Cambridge Philosophical Society* **69**, 95–115.

Peterson, C. J. & Facelli, J. M. (1992). Contrasting germination and seedling growth of *Betula allegheniensis* and *Rhus typhina* subjected to various amounts and types of plant litter. *American Journal of Botany* **79**, 1209–1216.

Reader, R. J. (1993). Control of seedling emergence by ground cover and seed predation in relation to seed size for some old-field species. *Journal of Ecology* **81**, 169–175.

Rees, M. (1995). EC-PC comparative analyses? *Journal of Ecology*, **83**, 891–892.

Rees, M. & Westoby, M. (1997). Evolution of seed size in a simple ecological model. *Oikos* **78**, 116–126.

Reich, P. B., Walters, M. B. & Ellsworth, D. S. (1992). Leaf life-span in relation to leaf, plant, and stand characteristics among diverse ecosystems. *Ecological Monographs* **62**, 365–392.

Roush, W. (1995). When rigor meets reality. *Science* **269**, 313–315.

Salisbury, E. J. (1942). *The reproductive capacity of plants*. London: G. Bell & Sons.

Saverimuttu, T. & Westoby, M. (1996a). Components of variation in seedling potential relative growth rate: phylogenetically independent contrasts. *Oecologia*, **105**, 281–285.

Saverimuttu, T. & Westoby, M. (1996b). Seedling survival under deep shade in relation to seed size: phylogenetically independent contrasts. *Journal of Ecology* **84**, 681–689.

Shipley, B., Keddy, P. A., Moore, D. R .J. & Lemkt, K. (1989). Regeneration and establishment strategies of emergent macrophytes. *Journal of Ecology* **77**, 1093–1110.

Shipley, B. & Peters, R. H. (1990). The allometry of seed weight and seedling relative growth rate. *Functional Ecology* **4**, 523–529.

Smith, C. C. & Fretwell, S. D. (1974). The optimal balance between size and number of offspring. *American Naturalist* **108**, 499–506.

Swanborough, P. & Westoby, M. (1996). Seedling relative growth rate and its components in relation to seed size: phylogenetically independent contrasts. *Functional Ecology* **10**, 176–184.

Thompson, K. & Baster, K. (1992). Establishment from seed of selected Umbelliferae in unmanaged grassland. *Functional Ecology* **6**, 346–352.

Tilman, D. (1990). Constraints and tradeoffs: toward a predictive theory of competition and succession. *Oikos* **58**, 3–15.

Tilman, D. (1994). Competition and biodiversity in spatially structured habitats. *Ecology* **75**, 2–16.

van der Werf, A., van Nuenen, M., Visser, A.J., & Lambers, H. (1993). Contribution of physiological and morphological plant traits to a species' competitive ability at high and low nitrogen supply. *Oecologia* **94**, 434–440.

Vasquez-Yanes S. C. & Orozco-Segovia, A. (1992). Effects of litter from a tropical rainforest on tree seed germination and establishment under controlled conditions. *Tree Physiology* **11**, 391–400.

Westoby, M., Jurado, E. & Leishman, M. (1992). Comparative evolutionary ecology of seed size. *Trends in Ecology and Evolution* **7**, 368–372.

Westoby, M., Leishman, M. R., & Lord, J. M. (1995a). On misinterpreting the 'phylogenetic correction'. *Journal of Ecology*, **83**, 531–534.

Westoby, M., Leishman, M. R., & Lord, J. M. (1995b). Further remarks on phylogenetic correction'. *Journal of Ecology*, **83**, 727–730.

Westoby, M., Leishman, M. R., & Lord, J. M. (1995c). Issues of interpretation after relating comparative datasets to phylogeny. *Journal of Ecology*, **83**, 892–893.

Westoby, M., Rice, B. & Howell, J. (1990). Seed size and plant-growth form as factors in dispersal spectra. *Ecology* **71**, 1307–1315.

Willson, M. F. (1983). *Plant reproductive ecology*. New York: Wiley.

9 • Packaging and provisioning in plant reproduction

D. Lawrence Venable

1 Introduction

For the last several decades, simple theoretical models explaining the partitioning of reproductive allocation in plants have existed. These include offspring size and number models (Smith & Fretwell 1974; Lloyd 1987; Venable 1992) and models of sex allocation (Charnov 1979; Charlesworth & Charlesworth 1981; Lloyd 1984). The empirical literature shows that botanical interests have grown beyond the scope of these simple frameworks. Plant ecologists recognize that, due to the modular construction of plants, seed number is a vector of hierarchical packaging decisions: the number of seeds per fruit, the number of fruits per infructescence, and the number of infructescences per plant. Each of these hierarchical components of seed number has its own evolutionary ecology and many are active areas of empirical pursuit. For example, seed number per fruit has important consequences for seed predation (Bradford & Smith 1977; Herrera 1984), seed dispersal (Augspurger 1986) and sib competition (Casper 1990; Casper et al. 1992). Likewise, the number of fruits per inflorescence may impact frugivore behavior by affecting either attractiveness or handling time involved in feeding (Schupp 1993).

When seed number is considered as a hierarchical packaging strategy, some of its components are intimately related to floral traits and it becomes more difficult to justify ignoring male function in cosexual plants. The number of fruits per inflorescence is influenced by the number of flowers per inflorescence, if only in that the number of fruits must be less than or equal to the number of flowers (unless one considers fruits that separate into separate units which take on the ecological functions normally pertaining to the whole fruit (e.g. *Crossosoma*). Similarly, the number of infructescences and inflorescences per plant are intimately biologically related. Thus, seed provisioning and packaging decisions impact flower size and number decisions through the rich frequency-dependent interplay of female and male function.

Male function is also a set of hierarchical provisioning and packaging strategies. How large should pollen grains be? There are interesting relationships between pollen size and style length (e.g. Williams & Rouse 1990; Kirk 1992) and pollen size and selfing (Barrett *et al.* this volume). Pollen grain number can be divided into the number of pollen grains per anther (or even per dosage in species with carefully controlled within-anther dispensing strategies; e.g. Buchmann *et al.* 1977), the number of anthers per flower, the number of flowers per inflorescence and the number of inflorescences per plant. Each of these levels has its complex ecology related to pollinator behaviours such as transport and consumption of pollen, or visitation frequencies and durations (Harder & Thomson 1989). For example, having more flowers per inflorescence sometimes leads to greater attractiveness and more visits per flower (Rodriguez Robles *et al.* 1992), but also more geitonogamy, the impact of which varies with the breeding system (Harder & Barrett 1995). More pollen grains per anther may make a flower more attractive to pollen-collecting insects and result in more or longer visits, but a declining fraction of the grains may be successfully transported to appropriate stigmas (Harder & Thomson 1989).

Selection operates simultaneously on the hierarchical components of reproductive design. In this chapter I develop theoretical tools for considering all of the components together. The model can be used to address the following questions: How does selection on different components of female and male fitness interact? Are there simple rules that explain such interactions? When is it safe to consider the evolutionary ecology of subcomponents of the reproductive design in isolation? What is being left out when we do?

2 Models

First I will consider seed size and number models from a hierarchical-packaging point of view. Then I will progressively add the complications of cosexuality, cost allometry, and interacting male and female reproductive components.

2.1 Seed size and number with hierarchical packaging

The standard Smith–Fretwell model is a convenient starting point because it is familiar (Figure 9.1). An often sigmoidal curve gives the set of feasible seed sizes and corresponding seed fitnesses for a plant species in a particular environment.

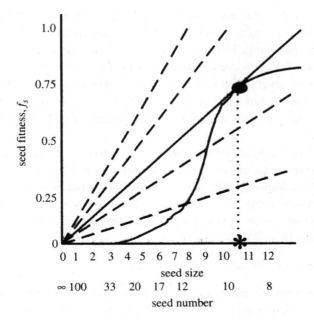

Figure 9.1. Graphical representation of the Smith–Fretwell model. The sigmoidal curve gives the set of feasible seed sizes and corresponding seed fitnesses for a plant species in a particular environment. For a given amount of resources, the seed-size axis implies a seed-number axis. Straight lines intersecting the origin represent lines of equal maternal fitness. For example, the second steepest dashed line passes through points with 10 seeds with fitness 1.0, 20 seeds with fitness 0.5, and 100 seeds with fitness 0.1. The steeper the straight line passing through the origin, the greater the maternal fitness. Thus the seed-size-number combination that maximizes maternal fitness in this particular environment is given by the point where an equal maternal-fitness line is tangent to the curve of feasible seed sizes and fitnesses (all other points on the curve of seed-size fitness have maternal-fitness lines with a shallower slope).

Algebraically, fitness for this model is described as $W(s,n) = nf(s)$ where $ns = R$. W equals maternal fitness (here a function of seed size and number), s equals seed size, $f(s)$ is the seed size fitness function for a plant species in a particular environment, graphed in Figure 9.1 as a sigmoidal curve, n equals seed number and R equals the resources available for making seeds. The first equation states that maternal fitness equals the number of seeds times the fitness per seed (which depends on seed size). The second equation explains how seed number is constrained by seed size and resources.

This model can be converted into a hierarchical provisioning and packag-

ing model if we redefine seed number to be the product of n (= number of
seeds per fruit), N (= number of fruits per inflorescence) and J (= number of
inflorescences per plant). The complex ecology of each hierarchical number
component can be considered by expressing the fitness of each by a potentially nonlinear general function: $W(s,n,N,J) = f_J(J)f_N(N)f_n(n)f_s(s)$ subject to
$JNns = R$. The evolutionary problem, as in the simple Smith–Fretwell
model, is to find the provisioning-packaging strategy that maximizes maternal fitness.

This can be done by maximizing $W(s,n,N,J)$, subject to the resource constraint, using the technique of Lagrange multipliers (see Chaing 1984; Lloyd
& Venable 1992; Venable 1992). This technique brings the resource constraint
into the function being maximized, in such a way as to insure that the
constraint is satisfied at the fitness maximum. The Lagrange function for the
above problem is:

$$L = f_J(J)f_N(N)f_n(n)f_s(s) - \lambda(JNns - R)$$

Notice that when the resource constraint is satisfied, the term to the right of
the first minus sign drops out and the Lagrange function is the same as the
fitness function. The technique involves finding the partial derivative of L
with respect to each packaging component, also treating λ as a variable,
setting each derivative equal to zero, finding the simultaneous solution and
checking for sufficiency conditions.

The partial derivative with respect to each variable has a similar structure
which can be illustrated with the partial derivative of J:

$$\frac{\partial L}{\partial J} = \frac{\partial f_J(J)}{\partial J}f_N(N)f_n(n)f_s(s) - \lambda Nns$$

$$= \frac{\partial f_J(J)}{\partial J}\frac{J}{f_J(J)}\frac{W(J,N,n,s)}{J} - \lambda Nns$$

$$= \frac{\partial \ln f_J(J)}{\partial \ln J}\frac{W(J,N,n,s)}{J} - \lambda Nns$$

Likewise for the other variables,

$$\frac{\partial L}{\partial N} = \frac{\partial \ln f_N(N)}{\partial \ln N}\frac{W(J,N,n,s)}{N} - \lambda Jns$$

$$\frac{\partial L}{\partial n} = \frac{\partial \ln f_n(n)}{\partial \ln n}\frac{W(J,N,n,s)}{n} - \lambda JNs$$

$$\frac{\partial L}{\partial s} = \frac{\partial \ln f_s(s)}{\partial \ln s} \frac{W(J,N,n,s)}{s} - \lambda\, JNn$$

$$\frac{\partial L}{\partial \lambda} = JNns - R$$

After setting each partial derivative equal to zero, the simultaneous solution is found by solving the first four equations for λ and setting them equal to each other:

$$\frac{\partial \ln f_J(J)}{\partial \ln J} = \frac{\partial \ln f_N(N)}{\partial \ln N} = \frac{\partial \ln f_n(n)}{\partial \ln n} = \frac{\partial \ln f_s(s)}{\partial \ln s} \tag{1}$$

$$J\,Nns = R$$

Thus the fitness-maximizing values of seed size, seed number per fruit, fruit number per inflorescence and inflorescence number per plant are found by equalizing the slopes of the component fitness functions with fitness functions and size and numbers expressed on log axes (cf. Venable 1992). This equality must further satisfy the condition that the product of the numbers and size equals the resources available for making seeds, as reflected in the last equation. The appropriate sufficiency condition to insure a fitness maximum rather than a minimum or saddle point is that the Hessian matrix is negative definite (Chaing 1984).

These partial derivatives on logarithmic scales are called 'elasticities' in economics and demography and they have intuitive conceptual meanings. Log changes can be thought of as proportional or percentage changes. Thus $\partial \ln f_s(s)/\partial \ln s$ can be thought of as a proportional or percentage change in per-seed fitness with a proportional or percent change in seed size. In economics, elasticities tell such things as how demand for a commodity will change with a change in price (both calculated as percentage changes). In demography, elasticities represent the proportional change in the population growth rate with a proportional change in a life-table parameter. As proportional changes, elasticities are independent of the scale on which a particular parameter is measured and also of the current values of the parameter.

Equation (1) tells us that selection operating on components of offspring size and number favours the equalization of the elasticities of the component fitness functions. There is no selection for change when the proportional increase in the fitness return from a size or number component due to a proportional increase in the magnitude of that component is equal for all size and number components. The logarithmic scale is natural due to the multipli-

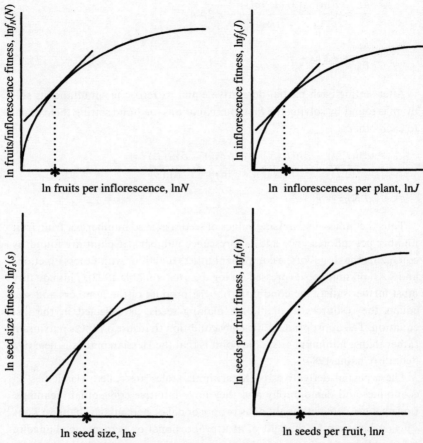

Figure 9.2. Graphical representation of the operation of selection on hierarchical provisioning and packaging strategies. Selection will favour reapportionment of the available resources among the seed size and number components until the slopes of the curves of log fitness components against log trait values are equal for all size and number traits (see equation (1)).

cative nature of the allocation constraint and of the fitness components.

The equal-elasticities solution reduces to the Smith–Fretwell model if fitness is a linear function of seed number and seed number is a single variable, i.e. if $W(s) = nf(s)$ where $ns = R$:

$$\frac{\partial \ln f_n(n)}{\partial \ln n} = \frac{\partial \ln f_s(s)}{\partial \ln s} \Rightarrow 1 = \frac{\partial \ln f_s(s)}{\partial \ln s} \Rightarrow 1 = \frac{\partial f_s(s)}{\partial s} \frac{s}{f_s(s)} \Rightarrow \frac{\partial f_s(s)}{\partial s} = \frac{f_s(s)}{s}$$

$$ns = R$$

This can be stated geometrically as: fitness is maximized where the slope for the seed-size-fitness function (the sigmoidal curve in Figure 9.1) equals the slope of a straight line passing through the origin, which is given by the value of the ordinate, $f_s(s)$, over the abscissa, s, for any point on the line (the dashed lines in Figure 9.1).

The operation of selection on hierarchical-provisioning and packaging decisions can be visualized graphically with Figure 9.2. Selection will favour reapportionment of the available resources among the seed-size and number components until the slopes of the curves in Figure 9.2 are equal. For simplicity and clarity of presentation, in the rest of this chapter I will assume that fitness components have graphs with the general shapes given in Figure 9.2. This is done to insure the sufficiency conditions for a fitness maximum. If a function curved upwards, equalizing its slope with that of other functions might result in minimizing fitness with respect to that fitness component. In such cases selection favors equalizing slopes at some other position on the graph where the slope curves downward or allocating at one of the boundaries of the x-axis (e.g. the largest or smallest possible seed sizes – coconuts and orchids?). The formal mathematical treatment of such cases is given in Lloyd & Venable (1992).

The most important implication of this model is that factors that determine or alter the shape of the three seed-number functions will alter the way selection operates on seed size and vice versa. Only if all three functions were linear, would several of the implications of the Smith–Fretwell model hold. These are that (1) the hierarchical partitioning of seed number will not affect fitness (alternative fruit designs and displays are selectively neutral) and (2) the seed-size-fitness function alone will determine the fitness-maximizing seed size (when resource availability varies, it is absorbed in seed number, not size).

2.2 Seed packaging and sex allocation

The number of fruits per infructescence and number of infructescences per plant will also be affected by factors impinging on floral ecology, since these are linked to floral traits through the level of fruit set. To explore these interactions for hermaphroditic plants requires explicit consideration of male function which may compete with female function for resources. Selection is frequency dependent because fitness through each sexual function depends not only on what an individual does, but also on the sexual strategies of its potential mates and sexual competitors.

Simple sex allocation models can be solved in terms of the elasticities of male and female fitness, which will be useful for integrating sex into hierarchical size-number problems. Let m equal allocation to male function, f equal allocation to female function, R equal total allocation, $f_\delta(m)$ equal male fitness before the operation of frequency-dependent interactions with the rest of the population (e.g. successful pollen production or removal), $f_\female(f)$ equal female fitness (e.g. seed production or successful seed dispersal). While female fitness of an individual can be considered to equal $f_\female(f')$ where the prime indicates the allocation decision of an individual in question, an individual's male fitness equals the total female fitness of the population times the proportion of these seeds sired by the individual in question. This equals $P f_\female(f) f_\delta(m')/P f_\delta(m)$ where P equals the population size of the mating group and the allocations without primes refer to those of the rest of the population (the individual in question is assumed to be a small fraction of the total population such that its strategy has an insignificant impact on the total female fitness and total male fitness in this expression). With the latter assumption (large population size), the Ps cancel so that the fitness of an individual with sexual allocations m' and f' in a population with sexual strategies m and f is given by

$$W(m',f',m,f) = f_\female(f') + \frac{f_\female(f) f_\delta(m')}{f_\delta(m)} \tag{2}$$

subject to the resource constraint $f' + m' = R$.

The Lagrange function is

$$L(m',f',m,f) = f_\female(f') + \frac{f_\female(f) f_\delta(m')}{f_\delta(m)} - \lambda(m' + f' - R)$$

The first step in finding the evolutionarily stable strategy (ESS) allocation to female and male function is to take the derivatives of the Lagrange function with respect to the individual's sexual strategy and the Lagrange multiplier (λ) and set the resulting expressions equal to zero:

$$\frac{\partial L}{\partial f'} = \frac{\partial f_\female(f')}{\partial f'} - \lambda = 0$$

$$\frac{\partial L}{\partial m'} = \frac{f_\female(f)}{f_\delta(m)} \frac{\partial f_\delta(m')}{\partial m'} - \lambda = 0$$

$$\frac{\partial L}{\partial \lambda} = f' + m' - R = 0$$

The ESS is found by solving these simultaneous equations and setting the individual sex allocation equal to the population sex allocation ($f' = f = \hat{f}$, $m' = m = \hat{m}$):

$$\frac{f_\female(\hat{f})}{\partial f} = \frac{f_\female(\hat{f})}{f_\male(\hat{m})}\frac{\partial f_\male(\hat{m})}{\partial \hat{m}} \Rightarrow \frac{\partial f_\female(\hat{f})}{\partial f}\frac{\hat{f}\hat{m}}{f_\female(\hat{f})} = \frac{\partial f_\male(\hat{m})}{\partial \hat{m}}\frac{\hat{m}\hat{f}}{f_\male(\hat{m})}$$

$$\Rightarrow \frac{\partial \ln f_\female(\hat{f})}{\partial \ln \hat{f}}\hat{m} = \frac{\partial \ln f_\male(\hat{m})}{\partial \ln \hat{m}}\hat{f}$$

$$\Rightarrow \frac{\dfrac{\partial \ln f_\female(\hat{f})}{\partial \ln f}}{\hat{f}} = \frac{\dfrac{\partial \ln f_\male(\hat{m})}{\partial \ln \hat{m}}}{\hat{m}}$$

$$\hat{f} + \hat{m} = R$$

At the ESS, female and male allocations are adjusted so that the fitness elasticity of female function divided by the female allocation equals the fitness elasticity of male function divided by male allocation and so that the resource constraint is satisfied. Rather than equalizing elasticities as in the hierarchical packaging and provisioning problem above, selection favours allocating to male and female functions in proportion to the fitness elasticities:

$$\frac{\hat{m}}{\hat{f}} = \frac{\dfrac{\partial \ln f_\male(\hat{m})}{\partial \ln \hat{m}}}{\dfrac{\partial \ln f_\female(\hat{f})}{\partial \ln \hat{f}}}$$

Sex allocation models often use power functions to describe male and female fitness components, e.g. $f_\female(f) = af^\alpha$; $f_\male(m) = bm^\beta$ (Lloyd 1984). The elasticities of power functions are the exponents. Thus, the ESS allocation ratio using these functions is $\beta : \alpha$ which can be interpreted in terms of the curvature of the male and female fitness functions (often called 'gain curves').

To convert such a model into a hierarchical provisioning and packaging model, we subdivide male and female allocation into hierarchical size and number components. Once again let seed size be s and seed number be the product of n (= number of seeds per fruit), N (= number of fruits per infructescence) and J (= number of infructescences per plant). The corresponding component fitness functions are still $f_s(s), f_n(n), f_N(N)$ and $f_J(J)$. The product of these will describe female fitness in an expanded equation (2). A variety of male packaging and provisioning fitness components can also be included in the model by letting p equal pollen grain size, w equal the number of pollen grains per stamen, x equal the number of stamens per flower, z equal the number of flowers per inflorescence and I equal the number of inflorescences per plant. In this initial model we will assume that selection operates separately on fruit and flower numbers per inflorescence or infructescence (subject to $N \leq z$), such that fruit set is simply the consequence of these separate selective outcomes. Likewise, we will initially assume that selection operates separately on infructescence and inflorescence numbers per plant (subject to $J \leq I$). The potentially nonlinear male fitness components are given by $f_p(p)$, $f_w(w)$, $f_x(x)$, $f_z(z)$ and $f_I(I)$. The resource constraint for this problem is $snNJ + pwxzI = R$ where the first product is total female allocation and the second is total male allocation. The equation for individual fitness is given by:

$$W(s', \ldots, J', p', \ldots, I', s, \ldots, J, p, \ldots, I)$$

$$= f_s(s')f_n(n')f_N(N')f_J(J')$$

$$+ \frac{f_s(s)f_n(n)f_N(N)f_J(J)f_p(p')f_w(w')f_x(x')f_z(z')f_I(I')}{f_p(p)f_w(w)f_x(x)f_z(z)f_I(I)}$$

The Lagrange function is

$$L(s', \ldots, J', p', \ldots, I', s, \ldots, J, p, \ldots, I)$$

$$= f_s(s')f_n(n')f_N(N')f_J(J')$$

$$+ \frac{f_s(s)f_n(n)f_N(N)f_J(J)f_p(p')f_w(w')f_x(x')f_z(z')f_I(I')}{f_p(p)f_w(w)f_x(x)f_z(z)f_I(I)}$$

$$- \lambda(s'n'N'J' + p'w'x'z'I' - R)$$

The ESS is found by finding the individual's hierarchical sex-allocation strategy that maximizes L for a given set of population parameters, then setting the individual's values of the reproductive-packaging components equal the population's values. The partial derivatives with respect to each component of an individual's reproductive strategy have similar structures which can be illustrated with the partial derivative with respect to s':

$$\frac{\partial L}{\partial s'} = \frac{\partial f_s(s')}{\partial s'} f_n(n')f_N(N')f_J(J') - \lambda n' N' J$$

$$= \frac{\partial f_s(s')}{\partial s'} \frac{s' \female tot'}{f_s(s')s'} - \lambda n'N'J'$$

$$= \frac{\partial \ln f_s(s')}{\partial \ln s'} \frac{\female tot'}{s'} - \lambda n'N'J'$$

where $\female tot'$ equals total female fitness of the individual in question $(= f_s(s')f_n(n')f_N(N')f_J(J'))$.

Similarly for the other variables

$$\frac{\partial L}{\partial n'} = \frac{\partial \ln f_n(n')}{\partial \ln n'} \frac{♀tot'}{n'} - \lambda s' N' J'$$

$$\frac{\partial L}{\partial N'} = \frac{\partial \ln f_N(N')}{\partial \ln N'} \frac{♀tot'}{N'} - \lambda s' n' J'$$

$$\frac{\partial L}{\partial J'} = \frac{\partial \ln f_J(J')}{\partial \ln J'} \frac{♀tot'}{J'} - \lambda s' n' N'$$

$$\frac{\partial L}{\partial p'} = \frac{\partial \ln f_p(p')}{\partial \ln p'} \frac{♂tot'}{p'} \frac{♀tot}{♂tot} - \lambda w' x' z' I'$$

$$\frac{\partial L}{\partial w'} = \frac{\partial \ln f_w(w')}{\partial \ln w'} \frac{♂tot'}{w'} \frac{♀tot}{♂tot} - \lambda p' x' z' I'$$

$$\frac{\partial L}{\partial x'} = \frac{\partial \ln f_x(x')}{\partial \ln x'} \frac{♂tot'}{x'} \frac{♀tot}{♂tot} - \lambda p' w' z' I'$$

$$\frac{\partial L}{\partial z'} = \frac{\partial \ln f_z(z')}{\partial \ln z'} \frac{♂tot'}{z'} \frac{♀tot}{♂tot} - \lambda p' w' x' I'$$

$$\frac{\partial L}{\partial I'} = \frac{\partial \ln f_I(I')}{\partial \ln I'} \frac{♂tot'}{I'} \frac{♀tot}{♂tot} - \lambda p' w' x' z'$$

$$\frac{\partial L}{\partial \lambda} = s' n' N' J' + p' w' x' z' I' - R$$

Setting all of these equations equal to zero, setting the individual packaging and provisioning strategy equal to the population strategy, recognizing that at the ESS, $♀tot = ♂tot$ and obtaining the simultaneous solution (by solving all but the last equation for λ), yields the following ESS condition:

$$\frac{\frac{\partial \ln f_s(\hat{s})}{\partial \ln \hat{s}}}{\hat{s}\hat{n}\hat{N}\hat{J}} = \frac{\frac{\partial \ln f_n(\hat{n})}{\partial \ln \hat{n}}}{\hat{s}\hat{n}\hat{N}\hat{J}} = \frac{\frac{\partial \ln f_N(\hat{N})}{\partial \ln \hat{N}}}{\hat{s}\hat{n}\hat{N}\hat{J}} = \frac{\frac{\partial \ln f_J(\hat{J})}{\partial \ln \hat{J}}}{\hat{s}\hat{n}\hat{N}\hat{J}}$$

$$= \frac{\frac{\partial \ln f_p(\hat{p})}{\partial \ln \hat{p}}}{\hat{p}\hat{w}\hat{x}\hat{z}\hat{I}} = \frac{\frac{\partial \ln f_w(\hat{w})}{\partial \ln \hat{w}}}{\hat{p}\hat{w}\hat{x}\hat{z}\hat{I}} = \frac{\frac{\partial \ln f_x(\hat{x})}{\partial \ln \hat{x}}}{\hat{p}\hat{w}\hat{x}\hat{z}\hat{I}} = \frac{\frac{\partial \ln f_z(\hat{z})}{\partial \ln \hat{z}}}{\hat{p}\hat{w}\hat{x}\hat{z}\hat{I}}$$

$$= \frac{\frac{\partial \ln f_I(\hat{I})}{\partial \ln \hat{I}}}{\hat{p}\hat{w}\hat{x}\hat{z}\hat{I}}$$

$$\hat{s}\hat{n}\hat{N}\hat{J} + \hat{p}\hat{w}\hat{x}\hat{z}\hat{I} = R$$

For this initial case, none of the female packaging/provisioning components is also a male packaging/provisioning component (or functions thereof) and vice versa. Under these conditions, the fitness elasticities for the female packaging/provisioning components will be equal to each other at the ESS just as in the hierarchical packaging model without sex. Likewise, male packaging/provisioning components will have equal elasticities at the ESS. The relationship between male and female elasticities satisfies $\female\hat{\varepsilon}/\hat{f} = \male\hat{\varepsilon}/\hat{m}$ subject to the resource constraint $\hat{f} + \hat{m} = R$, where $\female\hat{\varepsilon}$ and $\male\hat{\varepsilon}$ stand for female and male elasticities at the ESS and $\hat{f} = \hat{s}\hat{n}\hat{N}\hat{J}$ and $\hat{m} = \hat{p}\hat{w}\hat{x}\hat{z}\hat{I}$ are total female and total male allocations at the ESS (note that we need not specify which male and female elasticities, because they are all equal at the ESS). Thus under the conditions of this model the results of the separate hierarchical seed-size-and-number model and sex-allocation model are maintained: selection favours equilibration of size-number elasticities and favours allocation to male and female function in proportion to the male and female elasticities.

The interactions between male and female packaging strategies may be more complicated. For instance, the values of female provisioning and packaging components may constrain the values of male packaging components or vice versa. For example, fruit set could have some evolutionarily fixed value so that fruit number per infructescence and flower number per inflorescence constrain each other's evolution. Another issue that complicates the analysis is explicit consideration of accessory floral or fruit structures which represent indirect fitness components and may or may not impact both female and male reproductive success. Such factors include the production of nectar, petals and fruit dispersal structures.

The effects of such factors can be seen in the following model. Begin with the assumptions of the previous model, except let infructescence number per plant be constrained to equal inflorescence number ($J = I$). Let fruit number per inflorescence be $N = kz$, where fruit set, k, is a constant fraction of the number of flowers per inflorescence, z. Also let l be accessory floral costs (e.g. petals or nectar) that contribute to both male and female function and let d be fruit dispersal costs that contribute only to female fitness. The constraint on resource allocation is $snkzI + pwxzI + dkzI + lzI = R$, where the terms, from left to right, represent allocation to female size-and-number components, allocation to male size-and-number components, allocation to fruit dispersal and allocation to petals and nectar. The fitness function is:

$$W(s', \ldots, I', s, \ldots, I) = f_s(s')f_n(n')f_{\varphi l}(l')f_d(d')f_{\varphi z}(z')f_{\varphi I}(I')$$

$$+ \frac{f_s(s)f_n(n)f_{\varphi l}(l)f_d(d)f_{\varphi z}(z)f_{\varphi I}(I)f_p(p')f_w(w')f_x(x')f_{\Im l}(l')f_{\Im z}(z')f_I(I')}{f_p(p)f_w(w)f_x(x)f_{\Im l}(l)f_{\Im z}(z)f_{\Im I}(I)}$$

where $f_{\varphi l}(l)$ and $f_{\Im l}(l)$ represents the female and male fitness consequences of the level of allocation to petals or nectar and $f_d(d)$ represents the fitness consequences of the level of allocation to fruit dispersal structures. The same calculation procedures as for the previous model yield the following ESS conditions:

$$\frac{\dfrac{\partial \ln f_s(\hat{s})}{\partial \ln \hat{s}}}{\hat{s}\hat{n}k\hat{z}\hat{I}} = \frac{\dfrac{\partial \ln f_n(\hat{n})}{\partial \ln \hat{n}}}{\hat{s}\hat{n}k\hat{z}\hat{I}} = \frac{\dfrac{\partial \ln f_d(\hat{d})}{\partial \ln \hat{d}}}{\hat{d}k\hat{z}\hat{I}} = \frac{\dfrac{\partial \ln f_{\varphi l}(\hat{l})}{\partial \ln \hat{l}} + \dfrac{\partial \ln f_{\Im l}(\hat{l})}{\partial \ln \hat{l}}}{\hat{l}\hat{z}\hat{I}}$$

$$= \frac{\dfrac{\partial \ln f_{\varphi z}(z)}{\partial \ln \hat{z}} + \dfrac{\partial \ln f_{\Im z}(\hat{z})}{\partial \ln \hat{z}}}{(k(\hat{s}\hat{n} + \hat{d}) + \hat{p}\hat{w}\hat{x} + \hat{I})\hat{z}\hat{I}} = \frac{\dfrac{\partial \ln f_{\varphi I}(\hat{I})}{\partial \ln \hat{I}} + \dfrac{\partial \ln f_{\Im I}(\hat{I})}{\partial \ln \hat{I}}}{(k(\hat{s}\hat{n} + \hat{d}) + \hat{p}\hat{w}\hat{x} + \hat{I}) \, \hat{z}\hat{I}}$$

$$= \frac{\dfrac{\partial \ln f_p(\hat{p})}{\partial \ln \hat{p}}}{\hat{p}\hat{w}\hat{x}\hat{z}\hat{I}} = \frac{\dfrac{\partial \ln f_w(\hat{w})}{\partial \ln \hat{w}}}{\hat{p}\hat{w}\hat{x}\hat{z}\hat{I}} = \frac{\dfrac{\partial \ln f_x(\hat{x})}{\partial \ln \hat{x}}}{\hat{p}\hat{w}\hat{x}\hat{z}\hat{I}}$$

$$(\hat{s}\hat{n}k + \hat{d}k + \hat{p}\hat{w}\hat{x} + \hat{I})\hat{z}\hat{I} = R$$

Under the current assumptions, the fitness elasticities for the female packaging/provisioning components are no longer all equal to each other at the ESS. Likewise, male packaging/provisioning components do not all have equal elasticities at the ESS. Size–number components that impact only one gender function still have equal fitness elasticities at the ESS. Specifically, seed size and number per fruit have equal elasticities as do pollen grain size, pollen number per stamen and stamen number per flower. Also, size–number components that impact fitness of both genders have equal elasticities at the ESS, but it is the sum of their female and male elasticities that is equalized by selection. Thus, the sum of the female and male elasticities of the inflorescence number fitness component equals the sum of the fruits per inflorescence and flowers per inflorescence elasticities.

In the previous model, the denominators of the equal elasticity equation could be simply interpreted as total female or male allocation and the similarity to the simple non-hierarchical sex-allocation model was clear. More generally, the denominator in these expressions represent a scaling factor which converts a proportional change in a reproductive attribute into its total allocational cost. To calculate the total allocational cost associated

176 · D. L. Venable

with a shift in seed size, a given proportional change in seed size (e.g. 1%) is multiplied by the current seed size to give the cost per seed of this change. It is then further multiplied by the number of seeds per fruit, the number of fruits per inflorescence and the number of inflorescences per plant to get the full reproductive cost of a 1% change in seed size. Thus the ESS equality can be thought of as stating that the fitness elasticity of each fitness component will be equalized, when properly scaled to take into account the reproductive cost of a proportional change in each hierarchical size or number attribute. Because a 1% change in seed number per fruit has the same total reproductive cost as a 1% change in seed size, these costs cancel and the unscaled elasticities of seed size and seed number per fruit are equalized.

Some size-and-number attributes contribute to fitness through both male and female functions. The proportional fitness increase due to a proportional change in these attributes equals the sum of their female and male component elasticities (cf. terms for z and I in this model).

There are two fitness components in the model that do not represent male and female gamete packaging strategies, but rather accessory structure allocations. Since accessory structures do not have the same proportional costs as anything else, their denominators will not cancel.

The conclusion of this model is that equilibration of unscaled elasticities only occurs for proportional allocations that have the same proportional costs (e.g. the simple or hierarchical versions of the Smith–Fretwell model or the same-sex packaging components in the sexual model with no intersexual dependencies or accessory costs). Equilibration of elasticities for allocations with different reproductive costs requires scaling by the total reproductive cost of a proportional change in the component allocation.

2.3 The allometry of life history trade-offs

The constraint equations for the previous models (e.g. $sn = R$) are intuitive, but a plant functional morphologist would tell us that they are not likely to be empirically realistic. A ten-seeded fruit may cost more or less than ten times the production cost of a one-seeded fruit due to various fixed costs (economies of scale) or accelerating costs. Such cost non-linearities are frequently accounted for with allometric power equations, such that the resource cost of seed number per fruit, for example, might equal bn^β. Fixed costs imply $\beta < 1$ (a ten-seeded fruit costs less than ten times a one-seeded fruit) and accelerating costs are given by $\beta > 1$ (a ten-seeded fruit costs more than ten times a one-seeded fruit). Thus a more realistic constraint equation allowing cost nonlinearities for size-number components and accessory structures for the previous model is given by

$$(as^\alpha bn^\beta k + cd^\gamma\, k + ep^\varepsilon\, ow^\theta\, mx^\mu + ql^\pi)\, rz^\rho\, tl^\tau = R$$

where the Greek letter exponents are allometric constants translating reproductive attributes into their resource costs. For generality, I use general functions to translate reproductive attributes into their resource costs:

$$(g_s(s)g_n(n)k + g_d(d)k + g_p(p)g_w(w)g_x(x) + g_l(l))g_z(z)g_I(I) = R$$

When this constraint equation is used with the previous model, the ESS is found to be given by the following conditions:

$$
\frac{\dfrac{\partial \ln f_s(\hat{s})}{\partial \ln \hat{s}}}{\dfrac{\partial \ln g_s(\hat{s})}{\partial \ln \hat{s}}g_s(\hat{s})g_n(\hat{n})kg_z(\hat{z})g_I(\hat{I})} = \frac{\dfrac{\partial \ln f_n(\hat{n})}{\partial \ln \hat{n}}}{\dfrac{\partial \ln g_n(\hat{n})}{\partial \ln \hat{n}}g_s(\hat{s})g_n(\hat{n})kg_z(\hat{z})g_I(\hat{I})}
$$

$$
= \frac{\dfrac{\partial \ln f_d(\hat{d})}{\partial \ln \hat{d}}}{\dfrac{\partial \ln g_d(\hat{d})}{\partial \ln \hat{d}}g_d(\hat{d})kg_z(\hat{z})g_I(\hat{I})} = \frac{\dfrac{\partial \ln f_{\varphi l}(\hat{l})}{\partial \ln \hat{l}} + \dfrac{\partial \ln f_{\varphi l}(\hat{l})}{\partial \ln \hat{l}}}{\dfrac{\partial \ln g_l(\hat{l})}{\partial \ln \hat{l}}g_l(\hat{l})g_z(\hat{z})g_I(\hat{I})}
$$

$$
= \frac{\dfrac{\partial \ln f_{\varphi z}(\hat{z})}{\partial \ln \hat{z}} + \dfrac{\partial \ln f_{\varphi z}(\hat{z})}{\partial \ln \hat{z}}}{\dfrac{\partial \ln g_z(\hat{z})}{\partial \ln \hat{z}}(k(g_s(\hat{s})g_n(\hat{n}) + g_d(\hat{d})) + g_p(\hat{p})g_w(\hat{w})g_x(\hat{x}) + g_l(\hat{l}))g_z(\hat{z})g_I(\hat{I})}
$$

$$
= \frac{\dfrac{\partial \ln f_{\varphi I}(\hat{I})}{\partial \ln \hat{I}} + \dfrac{\partial \ln f_{\varphi I}(\hat{I})}{\partial \ln \hat{I}}}{\dfrac{\partial \ln g_I(\hat{I})}{\partial \ln \hat{I}}(k(g_s(\hat{s})g_n(\hat{n}) + g_d(\hat{d})) + g_p(\hat{p})g_w(\hat{w})g_x(\hat{x}) + g_l(\hat{l}))g_z(\hat{z})g_I(\hat{I})}
$$

$$
= \frac{\dfrac{\partial \ln f_p(\hat{p})}{\partial \ln \hat{p}}}{\dfrac{\partial \ln g_p(\hat{p})}{\partial \ln \hat{p}}g_p(\hat{p})g_w(\hat{w})g_x(\hat{x})g_z(\hat{z})g_I(\hat{I})} \tag{3}
$$

$$
= \frac{\dfrac{\partial \ln f_w(\hat{w})}{\partial \ln \hat{w}}}{\dfrac{\partial \ln g_w(\hat{w})}{\partial \ln \hat{w}}g_p(\hat{p})g_w(\hat{w})g_x(\hat{x})g_z(\hat{z})g_I(\hat{I})}
$$

$$= \frac{\dfrac{\partial \ln f_x(\hat{x})}{\partial \ln \hat{x}}}{\dfrac{\partial \ln g_x(\hat{x})}{\partial \ln \hat{x}} g_p(\hat{p}) g_w(\hat{w}) g_x(\hat{x}) g_z(\hat{z}) g_I(\hat{I})}$$

$$(g_s(\hat{s}) g_n(\hat{n})k + g_d(\hat{d})k + g_p(\hat{p}) g_w(\hat{w}) g_x(\hat{x}) + g_I(\hat{I})) g_z(\hat{z}) g_I(\hat{I}) = R$$

These ESS conditions are very similar to the previous ones: the elasticities of all male and female fitness components will be equilibrated at the ESS when properly scaled to take into account the reproductive cost of a proportional change in each hierarchical size or number attribute. The resource cost scaling factors still take into account the total reproductive cost of a proportional change in each hierarchical size or number attribute. But because costs are non-linear, the cost elasticity appears in the scaling factor. The resource-cost elasticity of a reproductive trait such as seed size is the proportional change in seed-size cost with a proportional change in seed size. This equalled one with the previous linear cost constraint. Thus, it dropped out of all previous ESS conditions. Now, a given proportional change in seed size (e.g. 1%) is multiplied by seed-size cost elasticity to get the proportional change in resource cost corresponding to the 1% change in seed size. This proportional change in resource costs is then multiplied by the current cost of a seed (given by the potentially nonlinear allometric cost function $g_s(s)$) to give the cost per seed of a proportional change for a seed of that size. The result is further multiplied by the costs associated with the number of seeds per fruit, the number of fruits per inflorescence and the number of inflorescences per plant to get the full reproductive cost of the 1% change in seed size.

One consequence of allometric constraints is that none of the reproductive traits will have equal unscaled fitness elasticities at the ESS unless their resource-constraint elasticities happen to be equal. More specifically, if the resource-cost elasticities are equal for the sets of traits that had equal unscaled elasticities in the previous model, selection will still favor equalizing their unscaled elasticities in the present model. All cost elasticities in the previous model equalled one, which satisfies this rule. For allometric resource constraints described by power functions, the elasticities are the exponents of the power functions. Thus the conclusions of the previous model would still hold in the present model if all reproductive traits had non-linear resource costs with the same allometric constant.

The impact of these allometric considerations is seen in the following example. In the previous non-allometric model, seed size and seed number per fruit were predicted to have equal unscaled elasticities at the ESS. Assume that seed-size and number-per-fruit cost allometries are described by power functions so that their cost elasticities equal α and β respectively. Assume that the seed-size and number-per-fruit fitness components have the same general shape as in Figure 9.2. Also, assume that seed size and number per fruit involve fixed costs so that α and $\beta < 1$ (thus a seed that is twice as big costs less than twice as much to produce and a ten-seeded fruit costs less than ten one-seeded fruits). If α equals β, selection will favour equalizing the unscaled fitness elasticities of seed size and number per fruit (see equation (3)). If α does not equal β, selection will favour an allocation pattern that results in the ratio of seed-size to seed-number fitness elasticities equaling α/β (this is derived from the equality of the first two terms in equation (3)). Thus if the fixed costs of seed size were greater than those for seed number ($\alpha < \beta$), at the ESS, the fitness elasticity of seed number per fruit will be larger than that of seed size. With reference to Figure 9.2, this implies larger seeds in fewer-seeded fruits than for the case of equal allometric constants. Thus in the presence of greater fixed costs for seed size than for seed number, selection favours greater allocation to the trait with greater fixed costs than would be predicted by equalizing fitness elasticities. When a cost is an accelerating function of allocation to a trait (i.e. when the cost elasticity is greater than one), selection will favour less allocation to it than would otherwise be the case.

Cost nonlinearities can be described in a variety of ways besides power functions. But the general result holds that greater fixed costs for a reproductive trait results in selection for more allocation to that trait than would otherwise be favoured.

2.4 Interacting packaging components of fitness

The previous models have assumed that the hierarchical reproductive traits have independent effects on fitness, and that their interactions operate through the trade-offs determined by the resource constraints. Thus total fitness could be presented as a product of the separate potentially non-linear fitness components and we can discuss and graph a seed size component of fitness independently of a seed number component (Figure 9.2).

However, it may frequently be the case that fitness components interact in ways that are not adequately accounted for by separate multiplicative fitness components. For example, the shape of the seed-size fitness function might be influenced by seed number per fruit in a fleshy endozoochorous species (e.g. total seed mass per fruit, rather than seed number and seed size separately, might determine interactions with frugivores). Likewise, the shape of the seed

180 · D. L. Venable

number fitness functions may be determined by seed size if predator satiation is a selective factor for more seeds per fruit and predators respond to total seed mass per fruit, per inflorescence, or per plant. Seed size and number may also affect the shape of the fitness curve for allocation to fruit dispersal. This could occur if more flesh were required to convince frugivores to swallow a larger seed mass, or if a bigger wing is required for wind to carry a larger seed mass the same distance. It is possible to conceive of ways that pollen size and numbers at different hierarchical levels interact and affect the fitness curve for petals or nectar. For example, the benefits of more nectar may saturate more rapidly with few pollen grains per anther than with many. It is more difficult to imagine how male packaging decisions affect the shapes of fitness curves for female traits and vice versa (e.g. how pollen grain size affects the shape of the seed size fitness function), though such interactions are also possible.

To explore such fitness interactions, I will assume the biology of the previous model except that female and male fitness components cannot be broken down into separate multiplicative fitness components. Assume that male packaging traits do not affect the shape of the female fitness curves and vice versa. Let the interacting female fitness components be given by the general function $f_\varphi(\cdot) = f_\varphi(s,n,l,z,I)$ and the interacting male fitness components be given by $f_\delta(\cdot) = f_\delta(p,w,x.z,I)$. Following the same procedures as above, the ESS conditions are:

$$
\frac{\dfrac{\partial \ln f_\varphi(\cdot)}{\partial \ln \hat{s}}}{\dfrac{\partial \ln g_s(\hat{s})}{\partial \ln \hat{s}}g_s(\hat{s})g_n(\hat{n})kg_z(\hat{z})g_I(\hat{I})} = \frac{\dfrac{\partial \ln f_\varphi(\cdot)}{\partial \ln \hat{n}}}{\dfrac{\partial \ln g_n(\hat{n})}{\partial \ln \hat{n}}g_s(\hat{s})g_n(\hat{n})kg_z(\hat{z})g_I(\hat{I})}
$$

$$
= \frac{\dfrac{\partial \ln f_\varphi(\cdot)}{\partial \ln \hat{d}}}{\dfrac{\partial \ln g_d(\hat{d})}{\partial \ln \hat{d}}g_d(\hat{d})kg_z(\hat{z})g_I(\hat{I})} = \frac{\dfrac{\partial \ln f_\varphi(\cdot)}{\partial \ln \hat{l}} + \dfrac{\partial \ln f_\delta(\cdot)}{\partial \ln \hat{l}}}{\dfrac{\partial \ln g_l(\hat{l})}{\partial \ln \hat{l}}g_l(\hat{l})g_z(\hat{z})g_I(\hat{I})}
$$

$$
= \frac{\dfrac{\partial \ln f_\varphi(\cdot)}{\partial \ln \hat{z}} + \dfrac{\partial \ln f_\delta(\cdot)}{\partial \ln \hat{z}}}{\dfrac{\partial \ln g_z(\hat{z})}{\partial \ln \hat{z}}(k(g_s(\hat{s})g_n(\hat{n}) + g_d(\hat{d})) + g_p(\hat{p})g_w(\hat{w})g_x(\hat{x}) + g_l(\hat{l}))g_z(\hat{z})g_I(\hat{I})}
$$

$$
= \frac{\dfrac{\partial \ln f_\varphi(\cdot)}{\partial \ln \hat{I}} + \dfrac{\partial \ln f_\delta(\cdot)}{\partial \ln \hat{I}}}{\dfrac{\partial \ln g_I(\hat{I})}{\partial \ln \hat{I}}(k(g_s(\hat{s})g_n(\hat{n}) + g_d(\hat{d})) + g_p(\hat{p})g_w(\hat{w})g_x(\hat{x}) + g_l(\hat{l}))g_z(\hat{z})g_I(\hat{I})}
$$

$$= \frac{\dfrac{\partial \ln f_{\varnothing}(\cdot)}{\partial \ln \hat{p}}}{\dfrac{\partial \ln g_p(\hat{p})}{\partial \ln \hat{p}} g_p(\hat{p}) g_w(\hat{w}) g_x(\hat{x}) g_z(\hat{z}) g_I(\hat{I})} = \frac{\dfrac{\partial \ln f_{\varnothing}(\cdot)}{\partial \ln \hat{w}}}{\dfrac{\partial \ln g_w(\hat{w})}{\partial \ln \hat{w}} g_p(\hat{p}) g_w(\hat{w}) g_x(\hat{x}) g_z(\hat{z}) g_I(\hat{I})}$$

$$= \frac{\dfrac{\partial \ln f_{\varnothing}(\cdot)}{\partial \ln \hat{x}}}{\dfrac{\partial \ln g_x(\hat{x})}{\partial \ln \hat{x}} g_p(\hat{p}) g_w(\hat{w}) g_x(\hat{x}) g_z(\hat{z}) g_I(\hat{I})}$$

$$(g_s(\hat{s}) g_n(\hat{n}) k + g_d(\hat{d}) k + g_p(\hat{p}) g_w(\hat{w}) g_x(\hat{x}) + g_I(\hat{I})) g_z(\hat{z}) g_I(\hat{I}) = R$$

These ESS conditions look very similar to the previous ones, but now the fitness elasticities are the proportional change in total female or male fitness components with a change in each reproductive attribute. This is a more general result since it includes the possibility of interacting fitness components.

These ESS conditions more closely approach a general equal-marginal-advantage equation that states that, when changes in traits are scaled to represent common units of resource, the marginal fitnesses of all traits are equal at the ESS (Lloyd & Venable 1992). For all of the models presented in this paper, the numerators are the elasticities of total fitness with respect to the individual reproductive attributes. However, these numerators reduce to varying degrees to something more simple and measurable. The specific results for the models in this chapter are that marginal fitnesses are measured logarithmically as elasticities (because of the multiplicative nature of size-number and sex allocation models). In the present model, the numerators reduce to the elasticities of male and female fitness (because of the assumption that these are independent). In the previous cases of independent fitness components for packaging attributes, the numerators reduce to the fitness elasticities of the individual provisioning components. Thus we trade specificity of predictions about more readily measured biological properties for generality as we make fewer restrictive assumptions and approach a general marginal-advantage model.

Likewise, all the denominators in the models of this chapter are the change in total resource cost with respect to a change in each provisioning component measured in its own units. This result can be derived from a general equal-marginal-advantage-equation (Venable & Lloyd 1992). The resource constraints have varying degrees of complexity such that the denominator is sometimes the product of the cost elasticity of a particular component multiplied by the total costs of that and all other components. In other cases it is simply the product of certain components (when the component cost elasticity equals 1), or the denominator may drop out entirely for models where the total cost elasticity is equal for all provisioning components. We could construct even more complex cost allometries for which the resource cost of any provisioning component depended on the values of the others. This would result in more general cost elasticities in the denominators.

3 Applications and extensions

The models developed here help to integrate our ideas of how the evolutionary ecologies of different reproductive components impact each other and provide general conceptual tools for specifying these interactions. For example, some types of inflorescences, such as spadices and capitulae, tend to have sessile fruits with little change in the configuration of the infructescence from that of the inflorescence (with a few obvious exceptions such as Mimosoideae). In such inflorescences, fruits and flowers tend to respond closely to each other's evolution (Ramirez & Berry 1995) and separate models of their floral or fruit biology would be incomplete. While the importance of selective interactions and constraints among different reproductive characters has been widely recognized (e.g. Primack 1987), there have not been effective conceptual tools for dealing with them simultaneously in the frequency-dependent setting of sexuality.

There are many specific applications of this model that, while not necessarily dealing with the integration of the whole reproductive deployment, can help with the design and interpretation of specific experimental or comparative studies of subsets of traits not treated by conventional models. Such specific applications usually require further development or reduction of the models.

As an example, consider the evolution of inflorescence size in *Asclepias*. The early studies of *Asclepias* inflorescence design were very influential in proposing the 'male function hypothesis' for the evolution of floral attraction (Willson & Rathcke 1974; Queller 1983). Subsequently a fairly large and somewhat confusing literature has developed regarding *Asclepias* inflorescence size and inflorescence size in general (references in Willson 1994; Wyatt & Broyles 1994). Some confusion exists about how to integrate selection operating on inflorescence size and number, taking into account potentially synergetic or competing gender functions that operate in the frequency-dependent population context of sexuality. Authors variously discuss female versus male fitness on a per-plant, per-inflorescence, or per-flower basis. Also, the nature of resource constraints is sometimes unclear with occasional discussions of the advantages for both gender functions of the production of more and larger inflorescences.

The issues can be clarified by applying the general framework presented here to the problem of the number of flowers per inflorescence, z, and the number of inflorescences per plant, I. Focusing on the female and male fitness consequences of these hierarchical packaging components from equation (3), we have:

$$\frac{\dfrac{\partial \ln f_{\female z}(\hat{z})}{\partial \ln \hat{z}} + \dfrac{\partial \ln f_{\male z}(\hat{z})}{\partial \ln \hat{z}}}{\dfrac{\partial \ln g_z(\hat{z})}{\partial \ln \hat{z}}} = \frac{\dfrac{\partial \ln f_{\female I}(\hat{I})}{\partial \ln \hat{I}} + \dfrac{\partial \ln f_{\male I}(\hat{I})}{\partial \ln \hat{I}}}{\dfrac{\partial \ln g_I(\hat{I})}{\partial \ln \hat{I}}}$$

$$g_z(\hat{z})g_I(\hat{I}) = R$$

This ESS condition states that, but for potential differences in allometry, selection favours equalizing the sum of male and female elasticities of inflorescence size and number, subject to the constraint on resources. Assuming the same allometric constant for inflorescence size and number, i.e., $rz^p t I^p = R$ (or ignoring allometry, i.e. $zI = Q$), this reduces to

$$\frac{\partial \ln f_{\female z}(\hat{z})}{\partial \ln \hat{z}} + \frac{\partial \ln f_{\male z}(\hat{z})}{\partial \ln \hat{z}} = \frac{\partial \ln f_{\female I}(\hat{I})}{\partial \ln \hat{I}} + \frac{\partial \ln f_{\male I}(\hat{I})}{\partial \ln \hat{I}}$$

subject to $zI = R/rt)^{1/p}$ or $zI = Q$, where $(R/rt)^{1/p}$ or Q is the total number of flowers per plant. If fitness can be assumed to be a linear function of inflorescence number such that the main nonlinearities occur with respect to inflorescence design, the ESS condition further reduces to

$$-\frac{\partial \ln h_{\female z}(\hat{z})}{\partial \hat{z}} = \frac{\partial \ln h_{\male z}(\hat{z})}{\partial \hat{z}} \tag{4}$$

where $h_{\female z}(z)$ and $h_{\male z}(z)$ are per–flower female and male inflorescence size fitness components. This simple ESS condition can be stated verbally as follows: selection favours a shift in the number of flowers per inflorescence until the percentage increase in male fitness per flower is cancelled by an equal but opposite percentage decrease in female fitness per flower. This simple expression combines, under one specific set of assumptions, the operation of selection on inflorescence size and number, taking into account both gender functions operating in the frequency-dependent population context of sexuality. It emphasizes per-flower fitness components and a balance between competing female and male function. If both female and male per-flower fitness components increased or decreased with inflorescence size (i.e. did not compete), this equality could not be met. Thus, the ESS would be to produce either one large inflorescence (with increasing per flower fitness components) or many single-flowered inflorescences (if both female and male per flower fitness components decreased with inflorescence size). Also, if $\ln h_{\varphi z}(z)$ declines and $\ln h_{\vartheta z}(z)$ increases with the number of flowers per inflorescence, but both functions curve upward, the equality given by equation (4) would be a fitness minimum and the ESS will be to produce either one large or many single-flowered inflorescences (whichever has the higher fitness). These caveats about the shapes of the fitness curve represent an informal treatment of the sufficiency conditions for the ESS equality.

Fishbien & Venable (1996) use equation (4) to evaluate the fitness curves obtained by measuring pollinia removal and deposition rates and fruit-initiation rates in experiments which manipulated umbel size and number in natural field populations of *Asclepias tuberosa*. The female and male per-flower fitness functions, $h_{\varphi z}(z)$ and $h_{\vartheta z}(z)$, have the general shapes illustrated in Figure 9.3. Equation (4) predicts that the ESS should be at roughly 12 flowers per umbel, which is very close to the population and species mean for *Asclepias tuberosa*. Thus the general model reduces to a form that permits rigorous thinking about inflorescence design and aids the interpretation of experimental data.

Other potential applications of the framework outlined here are not difficult to find. Harder & Thomson (1989) provided empirically-derived pollen-packaging fitness curves and evaluated the male-fitness consequences of changes in pollen packaging under a variety of ecological scenarios. In their discussion they recognize that packaging allometry and female function may modify the way selection operates on pollen packaging. Vonhof & Harder (1995) provide an analysis of within- and between-species allometry of pollen size and number for 21 species of legumes. These allometric relationships are directly interpretable as the allometric constraint functions in this chapter. The models presented above provide a framework for combining information such as that of Harder & Thomson (1989) with that of Vonhof & Harder (1995) to see how the pollen packaging predictions might change. The models also provide a general framework for the ideas developed by Schoen & Dubuc (1990) regarding allometric constraints on inflorescence construction and the implications of inflorescence size for geitonogamy in

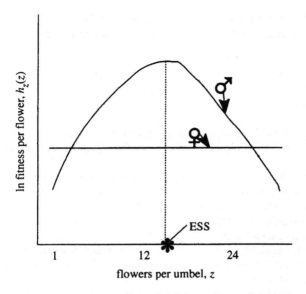

Figure 9.3. Graph of the general shapes of female and male per-flower fitness functions, obtained by measuring pollinia removal and deposition rates and fruit initiation rates in experiments which manipulated umbel size and number in natural field populations of *Asclepias tuberosa* (Fishbien & Venable 1996). Applying equation (4) to these data predicts that the ESS should be at roughly 12 flowers per umbel which is very close to the population and species mean for *Asclepias tuberosa*.

self-compatible and incompatible species. For example, they could be used to interpret the empirical results on the female and male fitness consequences of geitonogamy in Barrett *et al.* (1994) and Harder & Barrett (1995) under specific assumptions about the strength of inbreeding depression in *Eichhornia paniculata*. Perhaps the greatest utility of the model will be in such applications.

An important theoretical extension of the models presented here would be an analysis of the impacts of changing resource availability, *R*. Plants are highly plastic in size and shifts in the reproductive design with plant size have been predicted and observed (e.g., Venable 1992, Kang & Primack 1991). Furthermore, changes in the resource status of plants may change through the reproductive season, resulting in adaptive shifts in reproductive deployment (Kang & Primack 1991). Sectorial transport of resources in plants may also mean that adaptive reproductive deployments may vary among plants of the same size but differing degrees of resource integration. Since the resource constraint is explicitly considered in the modeling framework used here, the exploration of changes in ESS reproductive deployment with changes in resource availability would be straightforward (cf. Venable 1992).

References

Augspurger, C. K. (1986). Double- and single-seeded fruits of *Platypodium elegans*: consequences for wind dispersal and seedling growth and survival. *Biotropica* **18**, 45–50.
Barrett, S. C. H., Harder, L. D. & Cole, W. W. (1994). Effects of flower number and position on self-fertilization in experimental populations of *Eichhornia paniculata* (Pontederiaceae). *Functional Ecology* **8**, 526–535.
Bradford, D. F. & Smith, C. C. (1977). Seed predation and seed number in *Scheela* palm fruits. *Ecology* **58**, 667–673.
Buchmann, S. L., Jones, C. E. & Colin, L. J. (1977). Vibratile pollination of *Solanum douglasii* and *S. xanthi* (Solanaceae) in southern California. *Wasmann Journal of Biology* **35**, 1–25.
Casper, B. B. (1990). Seedling establishment from one- and two-seeded fruits of *Cryptantha flava*: a test of parent-offspring conflict. *American Naturalist* **136**, 167–177.
Casper, B. B., Heard, S. B. & Apanius, V. (1992). Ecological correlates of single-seededness in a woody tropical flora. *Oecologia* **90**, 212–217.
Chaing, A. C. (1984). *Fundamental methods of mathematical economics*. London: McGraw-Hill.
Charlesworth, D. & Charlesworth, B. (1981). Allocation of resources to male and female function in hermaphrodites. *Biological Journal of the Linnean Society* **15**, 57–74.
Charnov, E. L. (1979). Simultaneous hermaphroditism and sexual selection. *Proceedings of the National Academy of Sciences of the U.S.A.* **76**, 2480–2484.
Fishbien, M. & Venable, D. L. (1996). Evolution of inflorescence design: theory and data. *Evolution* **50**, 2165–2177.
Harder, L. D. & Barrett, S. C. H. (1995). Mating cost of large floral displays in hermaphrodite plants. *Nature, London* **373**, 512–515.
Harder, L. D. & Thomson, J. D. (1989). Evolutionary options for maximizing pollen dispersal of animal-pollinated plants. *American Naturalist* **133**, 323–344.
Herrera, C. M. (1984). Selective pressures on fruit seediness: differential predation of fly larvae on the fruits of *Berberis hispanica*. *Oikos* **42**, 166–170.
Kang, H. & Primack, R. B. (1991). Temporal variation of flower and fruit size in relation to seed yield in celandine poppy (*Chelidonium majus*; Papaveraceae). *American Journal of Botany* **78**, 711–722.
Kirk, W. D. J. (1992). Interspecific size and number variation in pollen grains and seeds. *Biological Journal of the Linnean Society* **49**, 239–248.
Lloyd, D. G. (1984). Gender allocation in outcrossing cosexual plants. In *Perspec-

tives on plant population ecology (ed. R. Dirzo & J. Sarukhán), pp. 277–300. Sunderland, Mass.: Sinauer Associates.

Lloyd, D. G. (1987). Benefits and costs of biparental and uniparental reproduction in plants. In *The evolution of sex: an examination of current ideas*. (ed. R. E. Michod & B. R. Levin), pp. 263–281. Sunderland, Mass.: Sinauer Associates.

Lloyd, D. G. & Venable, D. L. (1992). Some properties of natural selection with single and multiple constraints. *Theoretical Population Biology* **41**, 110–122.

Primack, R. B. (1987). Relationships among flowers, fruits, and seeds. *Annual Review of Ecology and Systematics* **18**, 409–430.

Queller, D. C. (1983). Sexual selection in a hermaphroditic plant. *Nature* **305**, 706–707.

Ramirez, N. & Berry, P. E. (1995). Production and cost of fruits and seeds in relation to the characteristic of inflorescence. *Biotropica* **27**, 190–205.

Rodriguez Robles, J. A., Melendez, E. J. & Ackerman, J. D. (1992). Effects of display size, flowering phenology, and nectar availability on effective visitation frequency in *Comparettia falcata*, Orchidaceae. *American Journal of Botany* **79**, 1009–1017.

Schoen, D. J. & Dubuc, M. (1990). The evolution of inflorescence size and number: a gamete-packaging strategy. *American Naturalist* **135**, 841–857.

Schupp, E. (1993). Opportunism vs. specialization: the evolution of dispersal strategies in fleshy-fruited plants. *Vegetatio* **107/108**, 107–120.

Smith, C. C. & Fretwell, S. (1974). The optimal balance between size and number of offspring. *American Naturalist* **108**, 499–506.

Venable, D. L. (1992). Size-number tradeoffs and the variation of seed size with plant resource status. *American Naturalist* **140**, 287–304.

Vonhof, M. J. & Harder, L. D. (1995). Size-number trade-offs and pollen production by papilionaceus legumes. *American Journal of Botany* **82**, 230–238.

Williams, E. G. & Rouse, J. L. (1990). Relationships of pollen size, pistil length and pollen tube growth rates in *Rhododendron* and their influence on hybridization. *Sexual Plant Reproduction* **3**, 7–17.

Willson, M. F. (1994). Sexual selection in plants: perspective and overview. *American Naturalists.* **144** (suppl.), S13–S39.

Willson, M. F. & Rathcke, B. J. (1974). Adaptive design of the floral display in *Asclepias syriaca*. *American Midland Naturalist* **92**, 47–57.

Wyatt, R. & Broyles, S. B. (1994). Ecology and evolution of reproduction in milkweeds. *Annual Review of Ecology and Systematics* **25**, 423–441.

IV • Recruitment and growth

10 · Comparative ecology of clonal plants

J.M. van Groenendael, L. Klimeš, J. Klimešová and R.J.J. Hendriks

1 Introduction

The specific capacity of plants to grow and reproduce clonally has had a continuous attraction for botanists. Detailed accounts of the morphology and anatomy of clonal plants exist from the middle of the last century onwards (Irmisch 1850; Velenovský 1907–1913; Goebel 1928–1933; Rauh 1937; Troll 1937–1942; Salisbury 1942; Leakey 1981). More recently research has focussed on the functionality of this fascinating complex of traits in adapting clonal plants to their environment (Harper 1977; Jackson *et al.* 1985; van Groenendael & de Kroon 1990; Callaghan *et al.* 1992; Soukupová *et al.*1994; Oborny & Podani 1996; de Kroon & van Groenendael 1997). Understanding the role of clonality in plants requires an integrated approach and cannot be achieved without understanding the morphological basis and its phylogenetic origin (Tiffney & Niklas 1985; Mogie & Hutchings 1990) that put boundaries to the functioning of clonal structures, and without knowledge of the growth rules that cause the specific spatial 'behaviour' of clonal plants (Bell 1991; Hutchings & de Kroon 1994; Oborny 1994). We equally require insight into the physiology of modularly organized systems (Watson & Casper 1984; Marshall 1990; Stuefer *et al.* 1994), that affect the direct performance of clonal organisms and we need to know the genetic (Ellstrand & Roose 1987; Eriksson 1993; Widén *et al.* 1994) and evolutionary (Tuomi & Vuorisalo 1989; Eriksson & Jerling 1990; Schmid 1990) consequences of asexual reproduction that affect the performance of clonal organisms in the long run.

Before we can start discussing constraints and mechanisms of the clonal growth form in the context of the functions of clonal growth, we need to define what constitutes a clonal plant. In the broadest sense, all plants are essentially clonal. Due to the late specialization of the plant cell, its so called long-preserved totipotency, and due to its modular construction where each module contains both somatic and meristematic tissue, all plants are potentially clonal and can be made to regenerate from plant parts. A reasonable working definition, however, is to describe a clonal plant as a plant capable of naturally producing potentially independent offspring by means of vegetative growth. This definition allows plants to be classified as clonal or non-clonal.

Broadly speaking, two approaches have been and still are followed to further our understanding of the role of clonality in adapting plants to their environment. The first approach, and also the approach that will be adopted in this chapter, is empirical and comparative and tries to achieve comprehensive descriptions of the patterns of distribution of clonality. The quality of such pattern analysis depends on both precise trait definition and precise habitat description; the analysis will ultimately produce inferences about the function of clonality, based on better insights in the various clonal growth forms and their interrelations. The possibilities for such pattern analysis are now increasing because of the greater availability of large sets of data on local floras that allow critical comparisons to be made.

What do we know so far about the patterns of distribution of clonal plant species over various habitats and over various plant communities? We know from field experience and published accounts that there exists an uneven distribution of clonal growth forms. Clonal growth is assumed to be more abundant in the arctic than in temperate regions (Perttula 1941; Callaghan *et al.* 1992), more abundant in aquatic habitats then in terrestrial ones (Duarte *et al.* 1994), more abundant in shady habitats like in forest understorey, and possibly more abundant under nutrient-limited conditions then under nutrient-rich conditions.

The second approach tries to unravel the working mechanisms in clonally organized plants and to demonstrate experimentally the costs and benefits involved in enhancing plant performance, working from hypotheses that functionally relate form and the patterns of distribution. The main functions ascribed to clonality can be divided into four categories: asexual recruitment, adult persistence, spatial mobility and storage.

1. It has been assumed, for instance, that the high risks involved in achieving pollination in submerged aquatic macrophytes have promoted clonal growth forms because these allow the production of asexual propagules (Duarte *et al.* 1994). Clonal fragmentation is a way to reduce the mortality risks of the genet and to produce an array of phenotypes from the same genotype (Eriksson & Jerling 1990) thus retarding the loss of genetic variation within populations (Silander 1985; Ellstrand & Roose 1987; Widén *et al.* 1994).

2. Low seedling recruitment in arctic habitats is compensated for by clonal persistence of established adults (Callaghan et al. 1992). Clonal persistence also allows the long-term occupation of a site, which is assumed to be an advantage when resources are scarce and need to be monopolized (de Kroon & Schieving 1990).

3. Through clonal growth, spatially distributed limiting resources can be exploited (Lovett Doust 1981; Alpert & Mooney 1986; Hutchings & de Kroon 1994; de Kroon & Hutchings 1995). It also enables growth in patchy environments, characterized by the simultaneous contrasts in more than one resource, achieving a certain division of labour (Stuefer et al. 1994, 1996; de Kroon et al. 1996).

4. Clonal plants can utilize stored reserves to tide them over difficult periods or give them a headstart at the beginning of a growing period (Shaver & Billings 1976; Klimeš et al. 1993; Geber et al. 1996). Not enough is known, however, about the costs and benefits of storage and remobilization of reserves so frequently stored in the clonal organs (Chapin et al. 1990).

Combining pattern and process in this functional way carries in itself two risks. The first one is that working mechanisms may be demonstrated for one species but may not be valid for others (Hutchings & de Kroon 1994; Kelly 1994). The second one and the one most relevant in the context of this chapter, is the assumption that pattern reflects function. The clonal traits that we are studying might not be the result of an optimization process such as current selection but the remnants of past evolutionary history (Gould & Lewontin 1979) or a byproduct of selection for a different trait, and this might seriously confound the functional relations we are trying to infer from a match between trait distribution and habitat distribution (Felsenstein 1985; Harvey et al. 1995; Westoby et al. 1995). Phylogeny is likely to be of importance, especially when dealing with clonal traits. As shown by Tiffney & Niklas (1985) and by Mogie & Hutchings (1990), the first vascular plants were clonal, but ancestors of the seed plants were woody and non-clonal, traits maintained among most gymnosperms. Also, the progenitor of the angiosperms is assumed to be non-clonal (Kelly 1995). Within the class of angiosperms the capacity for clonal growth became most prominent among the monocots that show reduction of secondary thickening and secondary monopolarity. Among dicots the pattern is more diversified. As pointed out by the above authors, it is in fact clonality that we need to consider as the derived trait among the seed plants.

The clonal growth characteristics that determine the performance of clonal plants under these various hypotheses of their function are basically few: the longevity of the functional connection between mother and daughter ramets; the distance between them; and the frequency and angle of branching. However, the morphological basis – the way a particular species realizes these traits – is much more varied and spans a range of morphologies, depending on the origin of meristems and their vascular connections within the clonal structures and the phyletic lineage of the species, each one imposing its own limitations on the ecological functions that we attribute to clonal growth.

What we set out to achieve in this study was to compare a large number of species with respect to their clonal characteristics and to relate these characteristics to the known habitat preferences of these species. More specifically we wanted (a) to explore the distribution of clonal traits over environmental gradients and to test the preference of clonal species for wetter, more resource poor and colder environments, and (b) to examine the assumed phylogenetic constraints in clonal traits and to test the degree to which the previously established trait–environment relationships are affected by a possible phylogenetic heritage.

2 Comparative analysis

To answer the questions raised above the flora of central Europe containing some 2300 species has been subjected to a comparative analysis. For each species the following data have been compiled:

1. Habitat preference based on indicator values according to Frank & Klotz (1990; largely adapted from Ellenberg (1991) to fit central European conditions). Indicator values range from 1 to 9 (and in the case of availability of water from 1 to 12) in an ordinal scale indicating a species preference in an environmental gradient. These gradients span habitats with low to high availability of light, nutrients, water or with low to high mean annual temperature.

2. Presence or absence of clonality, based on a comprehensive survey of the literature supplemented with extensive field observations, including morphological origin of the connection between shoots (root- or shoot-derived); longevity of connection in two categories (longer or shorter than one year); and lateral spread in cm also in two categories (longer or shorter then 25 cm). The more common clonal structures are categorized and presented in Figure 10.1. (For a more extensive description and finer

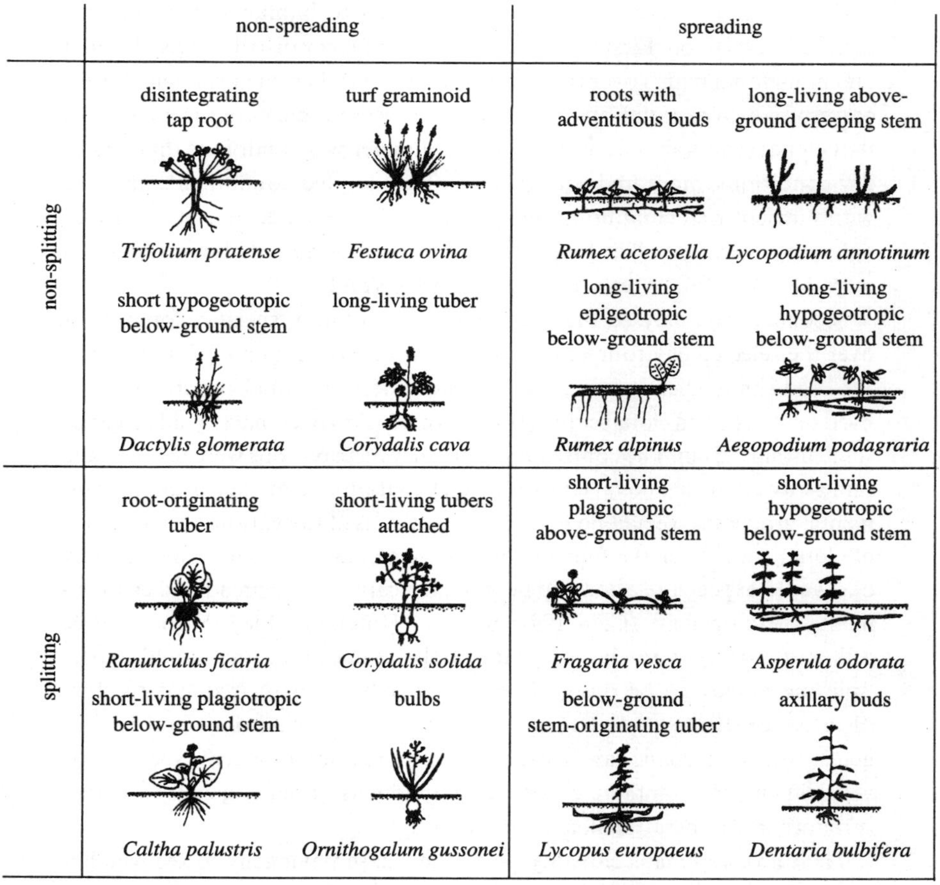

	non-spreading		spreading	
non-splitting	disintegrating tap root	turf graminoid	roots with adventitious buds	long-living above-ground creeping stem
	Trifolium pratense	*Festuca ovina*	*Rumex acetosella*	*Lycopodium annotinum*
	short hypogeotropic below-ground stem	long-living tuber	long-living epigeotropic below-ground stem	long-living hypogeotropic below-ground stem
	Dactylis glomerata	*Corydalis cava*	*Rumex alpinus*	*Aegopodium podagraria*
splitting	root-originating tuber	short-living tubers attached	short-living plagiotropic above-ground stem	short-living hypogeotropic below-ground stem
	Ranunculus ficaria	*Corydalis solida*	*Fragaria vesca*	*Asperula odorata*
	short-living plagiotropic below-ground stem	bulbs	below-ground stem-originating tuber	axillary buds
	Caltha palustris	*Ornithogalum gussonei*	*Lycopus europaeus*	*Dentaria bulbifera*

Figure 10.1. Various forms of clonal structures categorised according to their capacity to spread and the longevity of their connection. Origin of clonal structure and examples of species are given for each of the four categories.

details of clone morphology, not used in this chapter, see Klimeš *et al.* 1997).

3. The phylogenetic lineage of each species, established by placing each species in its family according to its current taxonomic position and relating families according to Chase *et al.* (1993), using one randomly selected tree from their second search and restricted to the plant families present in the central European flora.

On the species level we related clonal traits to habitat preference in a hierarchical fashion. First we contrasted the frequency distributions of clonal and non-clonal plants over the four major ecological gradients. Each of these corresponds to expectations based on larger geographic patterns, predicting that clonal plants should be more frequent under wet conditions than under dry conditions, more frequent under nutrient-limited conditions compared with nutrient-rich conditions, more frequent in the shade and at low mean annual temperatures. The frequency distributions were tested by straightforward Kruskal–Wallis tests (SAS release, SAS/STAT 1990).

Secondly, we analysed the frequency distribution of specific clonal traits over the selected environmental gradients within the group of clonal plants. We used simple dichotomies: root- or shoot-derived clonal structures; long-lived or short-lived clonal connection and extension of clonal spread larger or smaller than 25 cm. The contrasting frequency distributions were tested in the same way as those above. Unlike the distribution of clonality over environmental gradients, we had no *a priori* expectations of the various distributions of clonal traits over the four major environmental gradients. Theoretically one would expect a relationship between the capacity to spread and environmental heterogeneity (Cain 1994; Slade & Hutchings 1987; Sutherland & Stillman 1988). As to the longevity of the connection, one would expect 'splitting clones' to be more abundant in habitats with frequent random disturbance (Eriksson & Jerling 1990). However, as environmental heterogeneity and disturbance are usually defined in relation to specific species, it is very difficult to quantify environmental gradients of heterogeneity and disturbance in an operational, non-circular way.

The third step maps clonality on to the Chase phylogeny to see whether clonality shows a phylogenetic pattern that could affect the distribution patterns compiled from the database above. The Chase phylogeny is based on molecular data, independent from the morphological trait that is mapped onto it but, as the phylogeny is worked out at the family level only, the mapping is done for percentage occurrence of clonal species within a family. This introduces two possible sources of error. The first is the sometimes insecure assignment of species to higher order taxa, as this assignment is based on current, largely morphological criteria. The second is that some families, especially those with a centre of distribution outside central Europe, are not fairly represented by the species in the database. However, it is not very likely that these sources of bias affect clonal and non-clonal species differently.

In the last step the trait–environment relations were reanalysed on the level of the family while taking into account the phylogenetic descendance of such a taxon, using the method of independent contrasts (Harvey & Pagel 1991) and the CAIC algorithm (Purvis & Rambaut 1995). To reduce estimation errors in calculating family mean values of the relevant parameters, families containing less than five species were excluded. Contrasts were drawn with the independent environmental contrasts on the x-axis and the clonal trait contrasts on the y-axis and tested for significant relationships using regression through the origin.

3 Results

A simple comparison of the frequency distribution of the occurrence of clonality against the major environmental gradients shows a clear distinction between clonal and non-clonal species (Table 10.1). Clonal species are more frequent in wet habitats, under nutrient-poor conditions, at lower mean temperatures and also, although less clearly so, at lower light availability. This confirms, at the species level and at the scale of a regional flora, the qualitative observations from geographical patterns of distribution. It provides a strong basis for further extending functional theories to explain the observed trait–environment relationships.

Within the group of clonal plant species, clonality originates more frequently from stem structures compared to root-derived origins. The origin of a clonal structure is not randomly distributed over the major ecological gradients but sometimes shows clear preferences (Table 10.1). Quite obviously stem-derived clonal structures prevail in wet and aquatic habitats, but they are also more frequent under nutrient-rich conditions and under low light intensities. Fragmenting clones are less frequent than connected, persistent systems and show a preference for wet conditions and for nutrient-rich conditions (Table 10.1). Lastly, the capacity to spread is more evenly distributed over all clonal species but is also affected by habitat (Table 10.1) in more or less the same way as longevity of connection, with a preference for spreading in wet and in nutrient-rich habitats.

Looking at the mapping of the degree of clonality over the plant families present in the dataset, we can clearly observe a phylogenetic pattern (Figure 10.2). It is clear that there exists a relationship between monopolarity of the growth pattern, growth from one end only and the occurrence of, or rather necessity for, clonal growth. Within the monocots which as a group show secondary loss of bipolarity early during embryogenesis, the inferred frequency of clonal growth is apparently much higher than in the dicots, which

Table 10.1. *Medians of the frequency distribution of Ellenberg indicator values of (a) clonal and non-clonal plant species from the central European flora and (b) clonal species with respectively root or shoot derived clonal structures, short- or long-lived (> 1 yr) connections, and short- or long-distance (> 25 cm) capacity to spread.*

(Ellenberg values indicate low to high: availability of water (range 1–12); availability of nutrients (range 1–9); mean annual temperature (range 1–9); and availability of light (range 1–9); p values of Kruskal-Wallis test of difference in parentheses. In cases with significant differences between distributions but equal median values + and − signs indicate the direction of the difference.)

	Moisture		Nutrients		Temperature		Light	
	median	n	median	n	median	n	median	n
(a)								
Clonal	5	816	4	742	5	587	7−	851
Non-clonal	4	481	5	429	6	431	7+	536
	(0.0001)		(0.0001)		(0.0001)		(0.0001)	
(b)								
Root-derived	4	125	3	115	6	99	7+	138
Shoot-derived	6	699	4	610	5	477	7−	696
	(0.0001)		(0.016)		(0.0001)		(0.0007)	
Long-lived	5	628	4	592	5	469	7+	665
Short-lived	7	199	5	162	6	127	7−	199
	(0.0001)		(0.0001)		n.s.		(0.0001)	
Spreading < 25 cm	5	454	4	417	6	355	7	482
Spreading > 25 cm	7	334	5	297	5	213	7	338
	(0.0001)		(0.0015)		(0.0099)		n.s.	

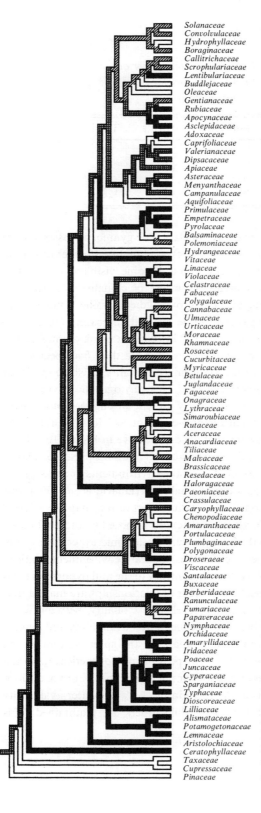

Solanaceae
Convolvulaceae
Hydrophyllaceae
Boraginaceae
Callitrichaceae
Scrophulariaceae
Lentibulariaceae
Buddlejaceae
Oleaceae
Gentianaceae
Rubiaceae
Apocynaceae
Asclepidaceae
Adoxaceae
Caprifoliaceae
Valerianaceae
Dipsacaceae
Apiaceae
Asteraceae
Menyanthaceae
Campanulaceae
Aquifoliaceae
Primulaceae
Empetraceae
Pyrolaceae
Balsaminaceae
Polemoniaceae
Hydrangeaceae
Vitaceae
Linaceae
Violaceae
Celastraceae
Fabaceae
Polygalaceae
Cannabaceae
Ulmaceae
Urticaceae
Moraceae
Rhamnaceae
Rosaceae
Cucurbitaceae
Myricaceae
Betulaceae
Juglandaceae
Fagaceae
Onagraceae
Lythraceae
Simaroubiaceae
Rutaceae
Aceraceae
Anacardiaceae
Tiliaceae
Malvaceae
Brassicaceae
Resedaceae
Haloragaceae
Paeoniaceae
Crassulaceae
Caryophyllaceae
Chenopodiaceae
Amaranthaceae
Portulacaceae
Plumbaginaceae
Polygonaceae
Droseraeae
Viscaceae
Santalaceae
Buxaceae
Berberidaceae
Ranunculaceae
Fumariaceae
Papaveraceae
Nymphaceae
Orchidaceae
Amaryllidaceae
Iridaceae
Poaceae
Juncaceae
Cyperaceae
Sparganiaceae
Typhaceae
Dioscoreaceae
Lilliaceae
Alismataceae
Potamogetonaceae
Lemnaceae
Aristolochiaceae
Ceratophyllaceae
Taxaceae
Cupressaceae
Pinaceae

Figure 10.2. Hypothetical reconstruction of the distribution of clonality over gymnosperm and angiosperm plant families from the central European flora, derived from weighted averages based on current frequency of clonality among extant plant families. Intensity of shading corresponds to increasing frequency of clonal species: solid denotes 75–100%; dark stippling denotes 50–75%; light stippling denotes 25–50%; open denotes 0–25%. Phylogeny based on Chase *et al.* (1993).

Table 10.2. *The distribution of clonality over genera represented by more than five species, for the subclasses of monocots and dicots*

	Monocots	Dicots	Total
Fully clonal	20	25	44
Mixed clonality	9	58	67
Fully non-clonal	0	14	15
Total	29	97	126

maintain their bipolarity in the entire subclass. The gymnosperms as a more ancient group are distinguished by their lack of clonality. Further down the phylogenetic tree, more phylogenetic effects become visible (Figure 10.2). Among the clonal monocot families some loss of clonality has occurred within the grass family. Among the large group of dicot families a much more diversified and polyphyletic pattern can be inferred, indicating the appearance and disappearance of clonal forms. This pattern is maintained at the lower level of genera. Among the genera with more than five species (126 genera in all) more then half of these show mixed clonality, i.e. some species in the genus are clonal and others are not. This phylogenetic trend is not restricted to dicot genera but is also clearly present among monocot genera (nine out of 29 contain both clonal and non-clonal growth forms; Table 10.2).

Given the phylogenetic pattern that we can observe in Figure 10.2, there are grounds to reanalyse the species–environment relations presented above. Especially when monocots dominate certain environments such as grasslands, sedge meadows and also aquatic habitats (Duarte *et al.* 1994), this could clearly affect general conclusions with respect to the adaptive value of clonality in these habitats. Given the restrictions imposed by the phylogeny, this reanalysis is confined to the family level. As one could expect by averaging the trait and habitat values over families, the trait–environment relationships become more robust. Nevertheless, contrary to the null hypothesis of no higher order phylogenetic effects, a number of relationships are maintained, notably the relation between nutrient levels and clonality, between water availability and clonality (Table 10.3), between water availability and longevity of connection, and between water availability and spreading, indicating higher level taxonomic effects and thereby possibly phylogenetic rather than ecological relationships. After correction for phylogenetic bias, out of the four strong relationships mentioned, the relationship between water availability and clonality disappears (Table 10.3), while others are maintained. This

Table 10.3. (a) *The relationship between percentage clonality and mean Ellenberg indicator values per family for four environmental gradients expressed as Pearson correlation coefficients and* (b) *independent contrasts based on the* CAIC *algorithm* (*Purvis & Rambaut* 1995) *calculated for percentage clonality and mean Ellenberg indicator values per family*

(The significance of the regression through the origin of the environmental gradient contrasts onto the clonality contrasts is given.)

	Moisture	Nutrients	Temperature	Light
(*a*)				
Correlation coefficient	0.31	-0.54	-0.30	0.08
n	48	47	47	48
Significance	$p < 0.0320$	$p < 0.0001$	$p < 0.0365$	n.s.
(*b*)				
n	47	46	46	47
Significance	$p < 0.12$	$p < 0.0008$	$P < 0.0211$	n.s.

implies that at least at the level of the family the relationship between water availability and clonality has a strong phylogenetic component.

4 Discussion

The comparative approach is one of the basic tools that provides the empirical basis for a scientific discipline, where observed patterns can be matched against theoretical expectations and used to generate hypotheses. Clearly, comparing species' traits and environmental characteristics is a way to achieve these goals in ecology. It is also clear that the criticism by Felsenstein and others on the comparative approach (e.g. Felsenstein 1985; Harvey *et al.* 1995) has something to contribute: there is a danger in considering species traits as independent measures of an ecological relationship as the actual relation is between individuals and their local environment. Inasmuch as the individual is part of a species it has a phyletic heritage constraining the traits and impacting on an individual's capacity to function. This criticism especially affects the evolutionary aspects of a species–environment relationship. Clearly, the direct effect of a trait on the functioning of an organism can be studied without paying much attention to the context into which such a trait is embedded: the breeding population of a species. It is when we cross the border between individual performance and individual fitness as most ecologists do, when we mix the proximate and the ultimate, that the phylogenetic constraints become an important aspect.

Having said that the comparative approach has to be applied with care when trying to explain trait–environment relationships as adaptations, because of the risk of phylogenetic constraints to be taken as adaptive traits, we immediately have to pose the reverse question. Should phylogenetically conservative traits always be treated as constraints? In other words, when we apply phylogenetic correction to our ecological comparisons of species, what is the risk that such a correction discards meaningfull trait–environment relationships just because they have happened to remain the same over evolutionary time? This has important methodological consequences. It implies that phylogenetic correction can support arguments but cannot disprove them.

Whenever a trait–environment relationship across species holds after taking into account possible effects of phyletic heritage, this provides strong corroboration of the adaptive value of such a trait, but the reverse is not necessarily true. As shown above, the relationship between clonality and the availability of water in a habitat disappears when taking independent contrasts. The probable reason for this is the fact that the majority of aquatic and semi-aquatic species are monocots, much more than the proportion of monocots in terrestrial habitats (Duarte *et al.* 1994). As pointed out by Duarte and his coworkers, plants have adapted to life in aquatic habitats from terrestrial ancestors. Given the specific problems of life under water, only a selected subset of available life forms actually succeeded, including a disproportionally large number of monocots. The main reason for this success, they assume, is the capacity of monocots to reproduce asexually and thereby overcome the problems involved with sexual reproduction in aquatic environments. If this is true, then clonality has a clear ecological function among aquatics that is now obscured by the phylogenetic correction, designed to discriminate between ecological and phylogenetic arguments.

The conclusion remains that a comparative analysis of trait–environment relationships based on a large set of species has confirmed the main intuitive patterns upon which a number of hypotheses as to the function of clonal traits have been built: clonal plants are more abundant under wet, cold and unproductive circumstances. Why clonal plants are more abundant under these circumstances is still largely hypothetical. One could argue that these three habitats all present abiotic limitations to the growing conditions for plants. These are species-poor habitats in general, and the species that do occur tend to be clonal. Clonality might be seen as a way of producing more feeding sites, conserving resources once acquired and providing a long life span by avoiding senescence as a way of overcoming sexual recruitment problems under unfavourable conditions.

Although this seems intuitively appealing it is not certain that such a general hypothesis for clonality is sufficient. One reason is the fact that among clonal plants themselves clonal traits show a different trend that seems to partially contradict the general idea of clonality as a way to survive cold, wet, resource-poor conditions. For instance, the tendency to spread, as well as the tendency to split, are more frequent under wet and but also under nutrient rich and shaded conditions, whereas more phalanx-like structures of tightly packed connected ramet systems occur under nutrient-poor but also dry and open non-shaded conditions.

This suggests in the first place that there is covariation in clonal traits, linking the longevity of the connection to the distance of spreading. This is supported by the significant negative correlation between frequency of long-lived connections at the family level against frequency of long-distance spreading (Pearson $R = -0.3163, p = 0.0303, n = 32$), although some of this relationship is lost when taking independent contrasts, probably because of the frequent occurrence of splitters among monocots and notably among the aquatics among them ($p = 0.0729$ for a regression through the origin).

This suggests in the second place that it might be the combination of traits that responds to combined nutrient, light and moisture gradients. This is supported by the relationship between mean Ellenberg values on the family level: preferences for nutrient-rich conditions covary negatively with the availability of light (Pearson $R = -0.4066, p = 0.0209, n = 32; p = 0.0038$ for the regression through the origin for independent contrasts) and positively, although less clearly so, with the availability of moisture (Pearson $R = 0.3112, p = 0.0829, n = 32; p = 0.0796$ for the regression through the origin for independent contrasts). This is represented schematically in Figure 10.3.

The preferred combinations of clonal traits are more abundant in the preferred combinations of environmental traits. Splitters (short-lived, long-distance connections) are preferably found under nutrient-rich and shaded or nutrient-rich and wet conditions, and the reverse is true for tightly packed, connected systems. This supports the idea that covariation in environmental characteristics forms an important aspect in trying to understand the functionality of clonal traits or trait combinations (Stuefer *et al.* 1996).

The costs and benefits of the trade-off that suggests itself here are as yet not quantified, but longer connections represent larger initial investments (de

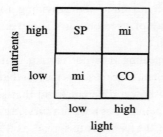 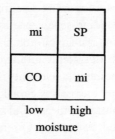

Figure 10.3. Environmental gradient space, showing in bold lining the significant negative association between nutrients and light ($R = -0.4066$; $p < 0.0209$; $n = 32$) and positive association between nutrients and water availability ($R = 0.3112$; $p < 0.0829$; $n = 32$). Covarying trait combinations ('clonal strategies') are mapped on this gradient space: short-lived, spreading connections (SPlitters) and long-lived, non-spreading connections (space COnsolidators). Other trait combinations are indicated as mixed (mi). Note the coincidence between covarying traits and habitat characteristics.

Kroon & Schieving 1990) and more costs of maintenance. These costs of maintenance get higher the longer a functional connection is maintained. Longer connections might represent a demographic benefit of dispersal, placing offspring well away from the parent. As soon as the propagule is formed, the connection is no longer functional and gets lost, leading to the splitting trait combination.

Longer connections also allow the exploitation of resources outside ramet depletion zones or when resources are patchily distributed. Here costs and benefits can be measured more directly. A substantial initial investment represents here a measurable risk and can be compared with the expectation of a return. Such risks are probably only worth taking in nutrient-rich habitats or in forest understorey where there are light gaps to be exploited. At the same time, a risky investment also invokes the need to restrict support when the investment is not returned, also leading to the trait combination of splitters. From a cost–benefit point of view, it will not always be easy to discriminate between the demographic benefit of asexual recruitment and the direct benefit of resource exploitation as the difference is mainly a question of timing. A new ramet can be produced more or less directly or after a phase of some sort of meristematic dormancy but in functional connection with the parent (Geber *et al.* 1996). However, when that connection gets lost before the ramet is produced, as in tuber or turion formation, the direct benefit is replaced by a longer-term demographic benefit that is much more difficult to quantify.

An important but unknown aspect in this cost–benefit balance is the extent to which an unsuccessful initial investment can be recovered, as already noted by Pitelka & Ashmun (1985). The fact is that long distance, short-lived connections (splitting clones) are relatively scarce among clonal plants (about 25%) and occur under wet, shaded and nutrient-rich conditions. They probably represent a derived rather than ancestral trait combination, contrary to the suggestion by Pitelka & Ashmun (1985).

Functional connections in a tightly packed system are relatively cheap and easy to maintain. They are indeed more common and mostly found under limiting growing conditions in dry and nutrient-poor habitats. It is unlikely that ramets well within each other's zone of influence could be made independent, and therefore exploitation of resources away from the parental depletion zone or asexual recruitment are unlikely explanations for the clonal trait combination of compact, long-lived tussock-like growth. Here the return on the investment might be the reduction of strong intraclonal competition which favours total clone persistence and site occupation, monopolizing the nutrients present (Carlsson & Callaghan 1990; Bullock *et al.* 1994). Another piece of evidence for covariation in persistence of connection and distance of spreading is the fact that within species, notably those with commercial value, such as *Festuca rubra* and *Trifolium repens*, breeding has created a wide array of clonal forms ranging from dense persistent tussock-like forms to strongly rhizomatous fragmenting forms (and in the case of *Trifolium* also non-clonal annual forms; N. R. Sackville-Hamilton, personal communication), indicating at least for these species the presence of the necessary genetic covariation that could explain the concerted appearance and disappearance of this clonal trait combination.

In conclusion, the large flora databases that are now becoming available are a rich source both for testing assumed ecological correlates, in this case trait–environment relationships in clonal plants, and exploring patterns that were not so obvious. The comparative method that generates these patterns should be used with caution. On the one hand it could lead to unwarranted relationships because constraints are taken for adaptations. On the other hand, the phylogenetic correction might point at an alternative explanation, but is not proof of such an alternative. Clonal plants are more frequent under wet, nutrient-poor and cold circumstances, as expected, but the functional relationships underlying such patterns are far less clear and require different hypotheses for different conditions. This is evident from the patterns of distribution of specific clonal traits, showing both covariation in traits and a preference of the trait combination to occur in specific combinations of environmental gradients. We need to improve our understanding of the partly conflicting functions assigned to clonal traits such as asexual reproduction versus resource exploitation, or persistence versus fragmentation, and this requires cost–benefit analyses of clonal growth with emphasis on recovery of costs already made.

We thank Jonathan Silvertown and Michael Dodd for their contribution to a phylogenetically correct analysis of our clonal data, and Joop Ouborg and Hans de Kroon for their comments on an earlier version of the manuscript. JvG gratefully acknowledges the financial support of the European Science Foundation. LK and JK were supported by grant A6005606 of the Academy of the Czech Republic.

References

Alpert, P. & Mooney, H. A. (1986). Resource sharing among ramets in the clonal herb *Fragaria chiloensis*. *Oecologia* **70**, 227–233.

Bell, A. D. (1991). *Plant form: an illustrated guide to flowering plant morphology*. Oxford University Press.

Bullock, J. M., Clear Hill, B. & Silvertown, J. (1994). Tiller dynamics of two grasses: responses to grazing, density and weather. *Journal of Ecology* **82**, 331–340.

Cain, M. L. (1994). Consequences of foraging in clonal plant species. *Ecology* **75**, 933–944.

Callaghan, T. V., Carlsson, B. A., Jonsdottir, I. S., Svensson, B. M. & Jonasson, S. (1992). Clonal plants and environmental change. *Oikos* **63**, 341–453.

Carlsson, B. A. & Callaghan, T. V. (1990). Programmed tiller differentiation, intraclonal density regulation and nutrient dynamics in *Carex bigelowii*. *Oikos* **58**, 219–230.

Chapin, F. S. III, Schulze, E.-D. & Mooney, H. A. (1990). The ecology and economics of storage in plants. *Annual Review of Ecology and Systematics* **21**, 423–447.

Chase, M. W.& 41 co-authors (1993). Phylogenetics of seed plants: an analysis of nucleotide sequences from plastid gene *rbc*L. *Annals of the Missouri Botanical Garden* **80**, 528–580.

Duarte, C. M., Planas, D. & Peñuelas, J. (1994). Macrophytes, taking control of an ancestral home. In *Limnology now: a paradigm of planetary problems* (ed. R. Margalef), pp. 59–79. Amsterdam: Elsevier.

Ellenberg, H. (1991). Indicator values for plants in central Europe. *Scripta Geobotanica* **18**, 1–248. (In German.)

Ellstrand, N. C. & Roose, M. J. (1987). Patterns of genotypic diversity in clonal plant species. *American Journal of Botany* **74**, 123–131.

Eriksson, O. & Jerling, L. (1990). Hierarchical selection and risk spreading in clonal plants. In *Clonal growth in plants: regulation and function* (ed. J. M. van

Groenendael & H. de Kroon), pp. 79–94. The Hague: SPB Academic Publishing.

Eriksson, O. (1993). Dynamics of genets in clonal plants. *Trends in Ecology and Evolution* **8**, 313–316.

Felsenstein, J. (1985). Phylogenies and the comparative method. *American Naturalist* **125**, 1–15.

Frank, D. & Klotz, S. (1990). Biological-ecological data to the flora of the GDR. *Wissenschaftliche Beitrage der Martin Luther Universität Halle-Wittemberg* **32**, 1–167. (In German.)

Geber, M., Watson, M. & de Kroon, H. (1996). Development and resource allocation in perennial plants: the significance of organ preformation. In *Plant resource allocation* (ed. F. A. Bazzaz & J. Grace), New York: Academic Press. (In the press.)

Goebel, K. (1928–1933). *Organographie der Pflanzen*, 3rd edn. Jena: Gustav Fischer. (In German.)

Gould, S. J. & Lewontin, R. C. (1979). The spandrels of San Marco and the Panglossian paradigm: a critique of the adaptationist programme. *Proceedings of the Royal Society of London* **B 205**, 581–598.

van Groenendael, J.M. & de Kroon, H. (1990). *Clonal growth in plants: regulation and function.* The Hague: SPB Academic Publishing.

Harper, J. L. (1977). *Population biology of plants.* London: Academic Press.

Harvey, P. H. & Pagel, M. (1991). *The evolutionary method in comparative biology.* Oxford University Press.

Harvey, P. H., Read, A. F. & Nee, S. (1995). Why ecologists need to be phylogenetically challenged. *Journal of Ecology* **83**, 535–536.

Hutchings, M. J. & de Kroon, H. (1994). Foraging in plants: the role of morphological plasticity in resource acquisition. *Advances in Ecological Research* **25**, 159–238.

Irmisch, T. (1850). *Zur Morphologie der monokotylische Knollen- und Zwiebelgewächse.* Berlin: G. Reimer Verlag. (In German.)

Jackson, J. B. C., Buss, L. W. & Cook, R. E. (1985). *Population biology and evolution in clonal organisms.* New Haven: Yale University Press.

Kelly, C. K. (1994). On the economics of plant growth: stolon length and ramet initiation in the parasitic clonal plant *Cuscuta europaea. Evolutionary Ecology*, **8**, 459–470.

Kelly, C. K. (1995). Thoughts on clonal integration: facing the evolutionary context. *Evolutionary Ecology* **9**, 575–585.

Klimeš, L., Klimešová, J., Hendriks, R. J. J. & van Groenendael, J. M. (1997). Clonal plant forms: classification, distribution and phylogeny. In *The ecology and evolution of clonal plants* (ed. H. de Kroon & J. M. van Groenendael). Leiden: Backhuys Publishers. (In the press.)

Klimeš, L., Klimešová, J. & Osbornová, J. (1993). Regeneration capacity and carbohydrate reserves in a clonal plant *Rumex alpinus*: effect of burial. *Vegetatio* **109**, 153–160.

de Kroon, H. & Schieving, F. (1990). Resource partitioning in relation to clonal growth strategy. In *Clonal growth in plants: regulation and function* (ed. J. M. van Groenendael & H. de Kroon), pp. 113–130. The Hague: SPB Academic Publishing.

de Kroon, H., Fransen, B., van Rheenen, J. W. A., van Dijk, A. & Kreulen, R. (1996). High levels of inter-ramet water translocation in two rhizomatous *Carex* species, as quantified by deuterium labelling. *Oecologia* **106**, 73–84.

de Kroon, H. & van Groenendael, J. M. (1997). *The ecology and evolution of clonal plants*. Leiden: Backhuys Publishers. (In the press.)

de Kroon, H. & Hutchings, M. J. (1995). Morphological plasticity in clonal plants: the foraging concept reconsidered. *Journal of Ecology* **83**, 143–152.

Leakey, R. R. B. (1981). Adaptive biology of vegetatively regenerating weeds. *Advances in Applied Biology* **6**, 57–90.

Lovett Doust, L. (1981). Population dynamics and local specialization in a clonal perennial (*Ranunculus repens*) I. The dynamics of ramets in contrasting habitats. *Journal of Ecology* **69**, 743–755.

Marshall, C. (1990). Source-sink relations of interconnected ramets. In *Clonal growth in plants: regulation and function* (ed. J. M. van Groenendael & H. de Kroon), pp. 23–41. The Hague: SPB Academic Publishing.

Mogie, M. & Hutchings, M. J. (1990). Phylogeny, ontogeny and clonal growth in vascular plants. In *Clonal growth in plants: regulation and function* (ed. J. M. van Groenendael & H. de Kroon), pp. 3–22. The Hague: SPB Academic Publishing.

Oborny, B. (1994). Growth rules in clonal plants and environmental predictability – a simulation study. *Journal of Ecology* **82**, 341–352.

Oborny, B. & Podani, J. (1996). *The role of clonality in plant communities*. Uppsala: Opulus Press. (In the press.)

Perttula, U. (1941). Studies on generative and vegetative reproduction of flowering plants from forest, meadow and rocky vegetation. *Annales Botanicae Societatis Zoologici-Botanici Fennici* Series A **58**, 1–388. (In German.)

Pitelka, L. F. & Ashmun, J. W. (1985). Physiology and integration of ramets in clonal plants. In *Population biology and evolution in clonal organisms* (ed. J. B. C. Jackson, L. W. Buss & R. E. Cook), pp. 399–435. Newhaven: Yale University Press.

Purvis, A. & Rambaut, A. (1995). Comparative analysis by independent contrasts (CAIC): an Apple Macintosh application for analysing comparative data. *Computer Applications in the Biosciences* **11**, 247–251.

Rauh, W. (1937). Die Bildung von Hypocotyl und Wurzelsprossen und ihre Bedeutung fuer die Wuchsformen der Pflanzen. *Nova Acta Leopoldina* **NF 4/24**, 395–552. (In German.)

Salisbury, E. J. (1942). *The reproductive capacity of plants*. London: Bell.

SAS/STAT (1990). *User's guide*, 4th edn. Cary NC: SAS Institute.

Schmid, B. (1990). Some ecological and evolutionary consequences of modular organization and clonal growth in plants. *Evolutionary Trends in Plants* **4**, 25–34.

Shaver, G. R. & Billings, W. D. (1976). Carbohydrate accumulation in tundra graminoid plants as a function of season and tissue age. *Flora* **165**, 247–267.

Silander, J. A. (1985). Microevolution in clonal plants. In *Population biology and evolution in clonal organisms* (ed. J. B. C. Jackson, L. W. Buss & R. E. Cook), pp. 107–152. New Haven: Yale University Press.

Slade, A. J. & Hutchings, M. J. (1987). The effects of light intensity on foraging in the clonal herb *Glechoma hederacea. Journal of Ecology* **75**, 639–650.

Soukopová, L., Marshall, C., Hara, T. & Herben, T. (1994). *Plant clonality: biology and diversity*. Uppsala: Opulus Press.

Stuefer, J. F., During, H. J. & de Kroon, H. (1994). High benefits of clonal integration in two stoloniferous species, in response to heterogeneous light environments. *Journal of Ecology* **82**, 511–518.

Stuefer, J. F., de Kroon, H. & During, H. J. (1996). Exploitation of environmental heterogeneity by spatial division of labour in a clonal plant. *Functional Ecology* **10**, 328–334.

Sutherland, W. J. & Stillman, R. A. (1988). The foraging tactics of plants. *Oikos* **52**, 239–244.

Tiffney, B. H. & Niklas, K. J. (1985). Clonal growth in land plants: a paleobotanical perspective. In *Population biology and evolution in clonal organisms* (ed. J. B. C. Jackson, L. W. Buss & R. E. Cook), pp. 35–66. New Haven: Yale University Press.

Troll, W. (1937–1942). *Vergleichende Morphologie der höheren Pflanzen, Band I–IV*. Berlin: Verlag Gebrüder Bornträger.

Tuomi, J. & Vuorisalo, T. (1989). Hierarchical selection in modular organisms. *Trends in Ecology and Evolution* **4**, 209–213.

Velenovský, J. (1907–1913). *Vergleichende Morphologie der Pflanzen Teil I-IV*. Prague: Fr. Řivnáč Verlag. (In German.)

Watson, M. A. & Casper, B. B. (1984). Morphogenetic constraints on patterns of carbon distribution in plants. *Annual Review of Ecology and Systematics* **15**, 233–258.

Westoby, M., Leishman, M. R. & Lord, J. M. (1995). On misinterpreting the 'phylogenetic correction'. *Journal of Ecology* **83**, 531–534.

Widén, B., Cronberg, N. & Widén, M. (1994). Genotypic diversity, molecular markers and spatial distribution of genets in clonal plants, a literature survey. In *Plant clonality: biology and diversity* (ed. L. Soukopová, C. Marshall, T. Hara & T. Herben), pp. 139–157. Uppsala: Opulus Press.

11 · Life history variation in plants: an exploration of the fast–slow continuum hypothesis

Miguel Franco and Jonathan Silvertown

1 Introduction

1.1 The comparative study of plant life histories

The comparative study of life histories has received considerable attention in the zoological literature (see Harvey & Pagel 1991, Stearns 1992). In plants, however, it has concentrated on traits related to reproductive systems and seed traits (e.g. Mazer 1990, Casper *et al.* 1992; Jordano 1995; Renner & Ricklefs 1995). The scarcity of studies of the whole life cycle of plants has meant that comparative studies of life history traits based on demographic information have mainly been attempted with closely related species (e.g. Sarukhán & Gadgil 1974; Fiedler 1987; Kawano *et al.* 1987; Fone 1989; Svensson *et al.* 1993). More than 20 years have gone by since the publication of Harper & White's (1974) *The demography of plants* and it is only now that we can attempt a comparative analysis of demographic traits based on a wider sample of detailed population studies.

A demographic classification of life histories

A usual approach in exploring the diversity of life histories is to look for patterns of correlation between life history and habitat. The existence of such correlation leads to the classification of life histories into a descriptive framework. Several authors (reviewed by Southwood 1988) have proposed classification schemes of life history variation. What these schemes have in common is the recognition of the role played by two environmental axes, productivity and disturbance.

Following this approach, we have suggested that elasticities may be useful in classifying species according to the relative importance of fecundity, stasis (survival without growth) and progression through the life cycle in different species and in different habitats (Silvertown *et al.* 1992, 1993). In a model of population dynamics where individuals are classified into n life stages and the demographic information is represented as contributions and transitions among all life stages in a square matrix ($\mathbf{A} = \{a_{ij}\}$ where i,j $= 1, 2, \ldots, n$), elasticity is defined as the proportional change in the finite rate of increase (λ_1) produced by a proportional change in each of the elements of the matrix (a_{ij}). That is, there exists a square matrix of elasticities ($\mathbf{E} = \{e_{ij}\}$) corresponding to the population matrix $\mathbf{A}$ such that $e_{ij} = \delta \ln(\lambda_1)/\delta \ln(a_{ij})$ (de Kroon et al. 1986).

Once the elasticity matrix has been obtained, the individual coefficients may be summed into identifiable demographic components. We have found it convenient to reduce the various demographic phenomena (progression to further stages of the life cycle, clonal growth, stasis or permanence in the same stage, retrogression to previous stages, seed production and seedling recruitment) into three main components: survival with positive growth (progression + clonal growth), survival without positive growth (stasis + retrogression) and fecundity (seed production and seedling recruitment). Because for each particular population matrix elasticities add up to unity (de Kroon et al. 1986; Mesterton-Gibbons 1993), species can be ordinated in a triangular space defined by these three demographic components (Figure 11.1; 'The demographic triangle'). Despite the fact that this ordination is limited by the mathematical constraints imposed upon the definition of elasticity itself (Shea, et al. 1994; Enright et al. 1995) and that it does not take into account the indirect effects that demographic parameters have on each other (Franco & Silvertown 1994; van Tienderen 1995), the triangular classification of plant populations is a first, heuristic step in framing the large variety of life histories.

Using an increasing sample of demographic information for perennial plant species has allowed us to conclude that the relative importance (elasticity) of these three population processes describes important dimensions of life history variation (Silvertown et al. 1992, 1993; Silvertown & Franco 1993). Specifically, the position of individual populations of the same species in the demographic triangle follows a gradient correlated both with successional stage (Figure 11.2, after Silvertown & Franco 1993) and with the population's rate of increase (Silvertown et al. 1996). Among species, this position is correlated with life form and habitat, but is also affected by longevity, relative growth rate and the number of life stages used to classify individuals in the population (Silvertown et al. 1993; Enright et al. 1995). Evidently, covariation among life history traits is largely responsible for the position that species or populations occupy in the triangle.

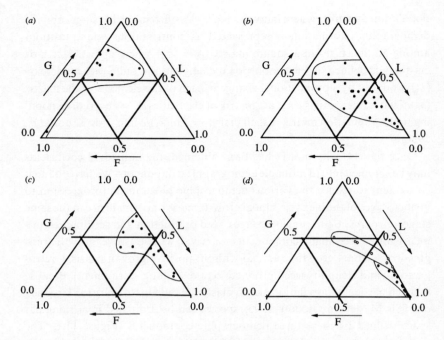

Figure 11.1. Ordination of species in the demographic triangle, updated from Silvertown *et al.* (1993). The relative importance (elasticity) of stasis (L), fecundity (F) and growth (G) derived from the analysis of population projection matrices is shown for different groups of species: (a) semelparous herbs, (b) iteroparous herbs of open habitats, (c) iteroparous forest herbs, (d) woody plants (closed circles denote trees; open circles denote shrubs).

The fast–slow continuum hypothesis of life history variation

The best known scheme of life history covariation is the r–K selection theory. Because of its emphasis on the relative roles of density-dependent and density-independent mortality, and its sometimes conflicting predictions, the theory has been repeatedly attacked and defended (see Boyce 1984). More recently, however, the emphasis of this discussion has moved from the kind of mortality (and habitat) encountered by organisms to the role played by mortality, whatever its cause, on other life history traits (e.g. Harvey & Zammuto 1985; Read & Harvey 1989; Promislow & Harvey 1990). In this respect, among other relationships Harvey and co-workers have found a negative relationship between adult mortality rate and age at sexual maturity, and a positive relationship between adult mortality rate and annual fecundity in mammals. This has led to the confirmation of a 'fast–slow' continuum, with

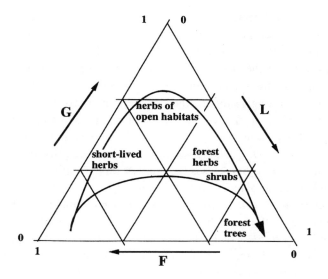

Figure 11.2. The successional trajectories suggested by the distribution of species in the demographic triangle in Figure 11.1, after Silvertown & Franco (1993).

organisms undergoing high adult mortality evolving fast development rates, high fecundity and short life cycles, on the one hand, and, on the other, organisms subject to low adult mortality developing slow growth rates, low fecundity and long life cycles.

Due to the inevitable covariation of demographic traits when populations are stationary (Sutherland *et al.* 1986), but also to the mathematical constraints imposed by the very definition of some of these traits, it becomes difficult to assign a unidirectional cause–effect relationship among them. Nevertheless, employing a combination of dimensional analysis, natural selection theory and stable demography, Charnov (1991) has derived a theoretical justification for the covariation of some of these demographic traits. In agreement with the evidence in animals (Harvey and co-workers' studies; Charnov & Berrigan 1990; Blackburn 1991; Charnov 1993; Berrigan *et al.* 1993; see also Roff 1992; Stearns 1992), Charnov's model predicts the signs found empirically in the relationships between adult mortality rate and age at sexual maturity (negative) and between adult mortality rate and fecundity (positive). This provides a theoretical justification of the fast-slow continuum which is free of the problems created by the interaction of external causes which have plagued previous models.

Charnov's model is based on three assumptions: (a) there is a trade-off between growth and reproduction, (b) once the adult stage is reached, mortality is constant and independent of reproduction; and (c) growth is determinate and annual fecundity is constant. While the second assumption defines the ecological setting, the other two refer to the organisms' morphological or physiological limitations and trade-offs. With these simple assumptions, Charnov's model strongly supports the idea that organisms can be ordinated between those living 'at a fast pace' (high mortality, short age at sexual maturity and lifespan, high fecundity), on one extreme, and those leading 'slower' ways of life (low mortality, long age at sexual maturity and lifespan, low fecundity), on the other. This is the simplest way of stating the fast–slow continuum hypothesis of life history variation.

In this chapter, we ask whether the fast–slow continuum hypothesis is applicable to plants. We show that a high rate of adult mortality is associated with a younger age at sexual maturity and shorter lifespan in genets of non-clonal plants and ramets of clonal plants. This is in agreement with the fast–slow continuum theory. However, because most plants monotonically increase their fecundity throughout adult life, the presumed positive relationship between mortality and fecundity cannot be tested. We therefore employ other demographic traits to explore the possible significance of this difference between unitary animals and modular plants.

2 Methods

2.1 Life history data

The dataset comes from published studies on plant demography. The original dataset containing information for 66 species is described by Silvertown *et al.* (1993). Eighteen more species were added (12 in Silvertown & Franco 1993 and a further six in Silvertown *et al.* 1997) and one deleted to give the 83 species used here. The paper for the deleted species (*Hypochoeris radicata*) did not contain the information necessary for the projection described below. The demographic information applies to genets of aclonal plants (53) but to ramets of clonal ones (30). Nine of the 83 species were semelparous. A more complete description of the dataset may be found in the papers cited above. An average matrix population model (see Caswell 1989) was projected for each of the 83 species using the program STAGECOACH (Cochran & Ellner 1992). This program yields a series of scalars, vectors and matrices useful in population studies, such as eigenvalue and eigenvector spectra, sensitivity and elasticity matrices, and several age-related life history parameters including survival and fecundity schedules. The parameters (traits) that we use in this chapter are:

1. r, the population's intrinsic rate of natural increase, calculated as the natural logarithm of λ_1, the dominant eigenvalue of each matrix;
2. L, total lifespan (the expected age at death for individuals that have already reached the last stage of the life cycle, equation (6) in Cochran & Ellner's paper);
3. α, age at sexual maturity (the expected age at which a newborn first enters the set of stages with positive fecundity; Cochran & Ellner's equation (15));
4. G, generation time (the mean age of parents of offspring produced at stable stage distribution; Cochran & Ellner's equation (26));
5. R_0, net reproductive rate (the total number of descendants in the lifetime of an average individual; Cochran & Ellner's equation (18)).

In cases where there was more than one type of recruit (e.g. seed, seedling and vegetative ramet) we used the population averages of these parameters provided by STAGECOACH.

The survival schedules generated by STAGECOACH were in turn used to calculate life expectancy at different ages. Because survival approaches zero asymptotically, the life cycle was assumed to end at the calculated average lifespan. Thus,

6. E_α, life expectancy at age at maturity, was used as an inverse estimate of adult mortality, M, $E_\alpha = 1/M$ (Harvey & Zammuto 1985; Sutherland *et al.*, 1986).

Finally, the life table survival and fecundity schedules were used to calculate the sensitivity of the intrinsic rate of natural increase to changes in age-specific survival and fecundity, employing the formulae given by Hamilton (1966; equations (1.6) and (1.7) in Rose 1991). These sensitivities measure the intensity of natural selection on these two parameters as the individuals age. The slope of the regression of the logarithm of each of these two sensitivities on age provided two scalars:

7. Hp_x, the exponential rate of decrease in the intensity of natural selection on survival:
8. Hm_x, the exponential rate of decrease in the intensity of natural selection on fecundity.

2.2 Taxonomic distribution and phylogeny of the species in the dataset

In order to account for the possible phylogenetic bias in our dataset we employed two classification systems. The first one corresponds to the taxonomic classification of Cronquist (Woodland 1991, after Cronquist 1981, for angiosperms, and Cronquist *et al.* 1966, for gymnosperms). In this system, the 83 species in our dataset were grouped into 70 genera, 39 families, 29 orders and 10 subclasses of angiosperms and gymnosperms. The second system used corresponds to the combined phylogeny of angiosperm (search II, Chase *et al.* 1993) and gymnosperm (Chaw *et al.* 1995) families obtained through the analysis of nucleotide sequences from the plastid gene *rbc*L and the 18S rRNA gene, respectively. The position of the Plantaginaceae, which was not present in the phylogeny of Chase *et al.*, next to the Scrophulariaceae was supported by both classical taxonomy (e.g. Cronquist, 1981) and the results of Olmstead & Reeves (1995) who sequenced the chloroplast genes *rbc*L and *ndh*F. The same justification was made for the Cactaceae, next to the Plumbaginaceae (among the 39 families in the dataset) within the Caryophyllids (*sensu* Chase *et al.* 1993), and the Capparaceae near the Cruciferae.

2.3 Data analyses

To identify the taxonomic level at which most variation occurs, a hierarchical analysis of variance was performed on all life history variables employing Cronquist classification. Furthermore, to incorporate the effect of taxonomic relatedness on the covariation of life history parameters, a comparative analysis by independent contrasts for each pair of relevant life history traits was conducted using each of the two classification systems. This was done with the help of the program CAIC (Purvis & Rambaut 1995).

The method, due to Felsenstein (1985), assumes a Brownian motion model of divergence among clades and, as a consequence, that the variance of the traits in question increases with the sum of branch lengths along the cladogram. Thus, given a phylogenetic tree, CAIC calculates a series of contrasts along the tree and, if information on branch lengths exists, the contrasts can be standardized by dividing them by the square root of the sum of those branch lengths, i.e. by their standard deviation (Felsenstein 1985, Garland 1992, Purvis & Rambaut 1995). If branch lengths are not available these can be assumed equal (an implicit view of punctuated evolution) or made proportional to the number of taxa below each node. The latter, however, requires that the full phylogeny be known.

Table 11.1. *Percentage of variance accounted for by different taxonomic levels in a hierarchical analysis of variance of life history traits employing Cronquist classification*

(For definition of traits, see text. For each trait, the highest value is represented in bold; second highest in italics.)

Life history trait	Taxonomic level						
	Division	Class	Subclass	Order	Family	Genus	Species
r	0.0	0.0	1.4	*36.4*	0.0	4.5	**57.7**
$\log(R_0)$	21.7	0.0	1.1	*33.4*	0.0	0.0	**43.9**
$\log(L)$	**69.4**	1.7	2.6	0.0	*10.4*	7.0	8.9
$\log(\alpha)$	**39.0**	0.0	16.0	0.0	*24.3*	5.0	15.7
$\log(G)$	**74.9**	0.0	4.9	0.0	4.8	6.7	*8.6*
$\log(E_z)$	**41.1**	0.6	2.6	15.4	0.0	*29.5*	10.8
Hp_x	0.0	0.0	14.1	*16.2*	0.0	**58.3**	11.3
Hm_x	0.0	3.8	0.0	**86.5**	0.0	*9.6*	0.0

In our comparative analyses, we assumed equal branch lengths. This is because in the case of the Cronquist system, having the same number of accepted taxonomic levels above each species implies equivalence within each level. In the other case, the phylogeny reconstructed from Chase *et al.* (1993) and Chaw *et al.* (1995) is based on two different genes with likely differences in their rates of evolution among lineages (e.g. Bousquet *et al.* 1992; Chase *et al.* 1993; Frascaria *et al.* 1993; Gunter *et al.* 1994). Furthermore, the reconstructed phylogeny for our dataset can only be applied down to the family level, after which the tree divides into as many genera and species as we had. Thus, the taxonomic diversity in our dataset is a measure of the diversity of authors' preferences.

3 Results

Three distinct patterns in the level where most of the variation is found can be distinguished in the life history parameters investigated (Table 11.1): (a) traits that are age-dependent (L, α, G and E_z) had their variation concentrated at the division level; (b) traits that can be described as time-dependent (r and R_0) had their variation concentrated at the species level; (c) traits that describe the rate of decrease in the intensity of natural selection with age (Hp_x and Hm_x) concentrated their variation at intermediate (genus and order, respectively) levels. In the first case, this is clearly a consequence of long-lived gymnosperms (*Sequoia*, *Araucaria* and *Pinus*) being contrasted with angiosperms

Table 11.2. *Pairwise correlations between life history traits taking individual species as independent points (lower diagonals) and as contrasts employing the Chase–Chaw phylogeny (upper diagonals)*

(In the latter analysis, the correlations on rows for r and $\log(R_0)$ correspond to these traits treated as *dependent*, while variables in rows $\log(L)$ to Hm_x were assumed *independent*. Values in parenthesis represent sample size, i.e. number of species and number of contrasts, respectively. Values in bold are significant at $P < 0.01$, usually $P < 0.001$; values in italics $0.01 < P < 0.05$.)

	r	$\log(R_0)$	$\log(L)$	$\log(\alpha)$	$\log(G)$	$\log(E_a)$	Hp_x	Hm_x
r		**0.63** (49)	−0.14 (53)	−0.09 (49)	−0.32 (52)	−0.16 (47)	**0.75** (31)	0.22 (31)
$\log(R_0)$	**0.73** (74)		0.11 (49)	0.01 (49)	−0.26 (48)	0.05 (47)	0.07 (29)	0.31 (29)
$\log(L)$	−0.21 (81)	0.01 (74)		**0.60** (49)	**0.68** (52)	**0.74** (47)	**−0.60** (31)	−0.04 (31)
$\log(\alpha)$	−0.19 (74)	−0.01 (74)	**0.70** (74)		**0.61** (48)	0.36 (47)	−0.38 (29)	*−0.37* (29)
$\log(G)$	**−0.45** (80)	**−0.34** (73)	**0.86** (80)	**0.75** (73)		**0.60** (46)	**−0.56** (31)	−0.25 (31)
$\log(E_a)$	−0.20 (64)	0.01 (64)	**0.90** (64)	**0.64** (64)	**0.78** (63)		**−0.50** (26)	−0.29 (26)
Hp_x	**0.84** (45)	0.34 (43)	**−0.54** (45)	**−0.46** (43)	**−0.67** (45)	**−0.54** (33)		0.06 (31)
Hm_x	0.26 (45)	0.21 (43)	−0.13 (45)	−0.22 (43)	−0.19 (45)	*−0.38* (33)	0.18 (45)	

which, although containing some trees, are dominated in the dataset by short-lived herbs. When calculating cross-species correlations, traits in the second group tended not to show covariation with those in the first group, except with G and Hp_x (Table 11.2). Of the two traits in the third group, Hp_x showed significant correlation with all traits, except Hm_x. The latter, on the other hand, only showed a weak correlation with E_α.

For the vast majority, these correlations remained the same when either taxonomy or phylogeny were incorporated into the analysis (Tables 11.2 and 11.3). Cross-species comparisons tended to yield higher slopes than comparisons using phylogeny, and these in turn gave higher slopes than comparisons employing taxonomy. As predicted, the relationship between life expectancy at maturity (E_α) and age at maturity (α), was positive and significant in cross-species ($P < 0.001$) and phylogenetic comparisons ($P < 0.05$).

4 Discussion

4.1 Fast–slow continuum hypothesis

The positive relationship between the inverse of adult mortality – life expectancy at maturity (E_α) – and age at sexual maturity (α) supports the hypothesis of a shorter life in species (or populations) that invest in sexual reproduction at an earlier age – or alternatively, that high risks of adult mortality favour those individuals which reproduce earlier (Figure 11.3, Table 11.3). This is also consistent with the positive relationship between age at maturity and adult lifespan found by Silvertown et al. (1997).

The inclusion of perennial plants among the groups of organisms on which these two relationships have been documented (see Stearns 1992; Charnov 1993) further supports the generality of the fast–slow continuum. It must be stressed, however, that Charnov's model of the fast–slow continuum was developed with mammals in mind and therefore his assumptions may not be totally appropriate for other organisms. As stated in the introduction, these assumptions are: (a) there is a trade-off between growth and reproduction, (b) once the adult stage is reached, mortality is constant and independent of reproduction; and (c) growth is determinate and annual fecundity is constant.

The first two assumptions are tenable in plants. There is ample evidence in the literature of a compromise between growth and reproduction in plants (e.g. Piñero et al. 1982; Bishop & Davy 1985; Eriksson 1985; Calvo & Horvitz 1990; Dick et al. 1990; Geber 1990; Pyke 1991; Gren & Willson 1994). Similarly, many species in our dataset show approximately constant mortality during adulthood. The third assumption, however, is simply not true for

Table 11.3. *Comparison of the regression results obtained between selected pairs of life history traits employing individual species data (cross-species comparison) or independent contrasts using Cronquist classification and the Chase–Chaw phylogeny*

(In the latter two cases, equal phylogenetic distances were assumed and the relationship was forced through the origin. Values in bold, $P < 0.001$; values in italics $0.01 < P < 0.05$.)

Independent	Dependent	Cross-species			Cronquist			Chase–Chaw		
		r	n	b	r	n	b	r	n	b
$\log(\alpha)$	$\log(E_z)$	**0.64**	64	0.721	n.s.	33		0.36	47	0.607
$\log(L)$	$\log(G)$	**0.86**	80	0.974	**0.67**	36	0.844	**0.68**	52	0.873
$\log(L)$	r	n.s.	81		n.s.	37		n.s.	53	
$\log(L)$	$\log(R_0)$	n.s.	74		n.s.	35		n.s.	49	
$\log(L)$	Hp_x	**-0.54**	45	-0.169	**-0.56**	22	-0.195	**-0.60**	31	-0.194
$\log(L)$	Hm_x	n.s.	45		n.s.	22		n.s.	31	

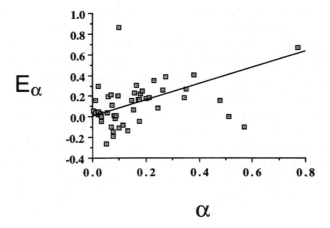

Figure 11.3. The relationship between life expectancy at maturity (E_α) and age at maturity (α), employing independent contrasts on the Chase–Chaw phylogeny $(n = 47; r = 0.36; P < 0.05)$.

the majority of plants. With the exception of plants with a single terminal meristem, which maintain a relatively constant annual fecundity during the adult stage, most plants show indeterminate growth through the iteration of multiple terminal meristems and, consequently, increasing fecundity with age (Watkinson & White 1985). Nevertheless, and in spite of the fact that plants violate the third assumption of Charnov's model, our data still conform to the expected negative correlation between adult mortality $(1/E)$ and age at maturity (α). However, because of their increasing fecundity with age/size/longevity, the expected positive relationship between mortality and fecundity cannot be tested.

In this respect, it is interesting to note that, contrary to what is observed in unitary animals, neither the intrinsic rate of population increase (r) nor the net reproductive rate (R_0) show the expected, significant negative correlations with longevity (L), age at maturity (α) and life expectancy at maturity (E_α), but, as ought to be expected, they do show correlation with generation time (G) (Table 11.2). The latter correlation, however, may be a trivial one as it is subject to the mathematical constraint of the definition of the traits themselves. Nevertheless, the result that the average rate at which the intensity of natural selection on survival diminishes with age (Hp_x) decreases with the lifespan of the species (L), and is correlated with all the other traits investigated, but, the same rate measured on fecundity (Hm_x) shows no relationship with all these variables (Table 11.2) adds to the evidence that the evolution of the reproductive schedules of plants does not have the same kind of restrictions imposed upon unitary organisms with a ceiling fecundity.

It is important to recognize that r and R_0 (but not Hm_x) could have any value between zero and the maximum attainable under ideal conditions. Earlier workers clearly understood this and compared the maximum attainable rate (r_{max}) of different species under ideal conditions (e.g. Evans & Smith 1952) or something approaching this under field conditions (e.g. Ross 1992). Until this information is available, we cannot be certain that the lack of correlation observed between these two traits and others in Table 11.2 is real or not. Nevertheless, the lack of relationship between Hm_x and all other parameters, coupled with the correlation between either r or R_0 and G, makes us feel confident that the effects are real and not a consequence of r and R_0 being labile.

In the case of animals with ceiling fecundity and decreasing reproductive value during adulthood, the trade-off between size and number of offspring implies that animals that are large at birth can only be produced in small numbers. The negative relationship between body size and fecundity and between body size and generation time implies that large animals will have smaller r than small, short-living animals.

Although with notable exceptions (lilies in temperate forests, the pioneer tropical tree *Cecropia obtusifolia*), the demography of plants has tended to concentrate on invasive short-lived herbs that, because of the absence of density-dependence and interspecific competition, have large r, and on long-lived trees, inhabiting stable communities, with $r \approx 0$. Despite this, the lack of relationship between r and L is remarkable and reinforces the view that the evolution of reproductive schedules does not follow the same rules in plants and animals. This is in agreement with the simulation study of Sackville-Hamilton *et al.* (1987) of a diversity of life histories evolving under similar environmental conditions, and of similar life histories evolving in different environments, depending on the architectural organization and allocation schedules between growth and reproduction of plants.

4.2 Cross-species vs comparative analyses

Recent comparative studies have found contrasting results to those employing cross-species analyses (e.g. Kelly & Purvis 1993; Kelly 1995a; Kelly & Beerling 1995). We believe that the similar results obtained here with the three schemes employed (cross-species, Cronquist classification and Chase–Chaw phylogeny) simply points to the limited scope of our dataset. For example, with the exception of the Alismatidae (reflecting the lack of demographic studies of aquatic plants; Franco & Silvertown 1990), all subclasses of angiosperms are represented in the dataset, but there is at present no way of knowing if the variation found at lower taxonomic levels is representative of the subclass as a whole. Most likely it is not. Having 83 species distributed in 39 families means that each family is, on average, represented by two species. Similarly, with 39 families and 29 orders, most orders are represented by a single family. Whether or not taxonomic relatedness affects the results of a comparative analysis depends on the completeness/diversity of the dataset (and obviously on the phylogeny employed; Coddington 1992) but, to make things worse, this possibly may also depend on the kind of information being analysed (Kelly & Woodward, this issue). A wider dataset may be necessary before we can work our way down to investigate the covariation of demographic traits at lower taxonomic levels, particularly if we want to investigate the effect of categorical variables like habitat, life form and parity.

4.3 The future of comparative studies

We have only scratched the surface of life history variation in plants. Despite differences in the reproductive schedules of unitary animals and most perennial plants, the similarities found in the covariation of life history traits are consistent with theoretical expectations. Those differences, and the way they may affect other traits not considered here, are worthy of further investigation. Certainly, those differences have to do not only with the modular construction of plants but with the hierarchical organization of the whole plant body. For example, it would be interesting to investigate how different degrees of morphological and physiological integration (see Kelly 1995b), different degrees of branching, and different degrees of rooting capabilities affect the covariation of demographic life history traits.

We wish to thank Rubén Pérez-Ishiwara for his skillful support in the technical aspects of the project. Several people contributed with their advice, direct help, and discussion, among them Elena Alvarez-Buylla, Mark W. Chase, Fernando Chiang, Mike Dodd, Luis Eguiarte, Colleen K. Kelly, Kevin McConway, Daniel Piñero and Larry Venable. Our collaboration has been aided by The British Council, The Royal Society, The Academia de la Investigación Científica, The Dirección General de Asuntos del Personal Académico, UNAM (Project IN209893) and CONACyT, México.

References

Berrigan, D., Charnov, E. L., Purvis, A. & Harvey, P. H. (1993). Phylogenetic contrasts and the evolution of mammalian life histories. *Evolutionary Ecology* **7**, 270–278.

Bishop, G. F. & Davy, A. J. (1985). Density and the commitment of apical meristems to clonal growth and reproduction in *Hieracium pilosella*. *Oecologia* **66**, 417–422.

Blackburn, T. M. (1991). Evidence for a 'fast-slow' continuum of life-history traits among parasitoid Hymenoptera. *Functional Ecology* **5**, 65–74.

Bousquet, J., Strauss, S. H. & Li, P. (1992). Complete congruence between morphological and *rbc*L-based phylogenies in birches and related species (Betulaceae). *Molecular Biology and Evolution* **9**, 1076–1088.

Boyce, M. S. (1984). Restitution of *r*- and *K*-selection as a model of density-dependent natural selection. *Annual Review of Ecology and Systematics* **15**, 427–448.

Calvo, R. N. & Horvitz, C. C. (1990). Pollinator limitation, cost of reproduction, and fitness in plants: a transition-matrix demographic approach. *American Naturalist* **136**, 499–516.

Casper, B. B., Heard, S. B. & Apanius, V. (1992). Ecological correlates of single-seededness in a woody tropical flora. *Oecologia* **90**, 212–217.

Caswell, H. (1989). *Matrix population models*. Sunderland, MA: Sinauer.

Charnov, E. L. (1991). Evolution of life history variation among female mammals. *Proceedings of the National Academy of Sciences of the U.S.A.* **88**, 1134–1137.

Charnov, E. L. (1993). *Life history invariants*. Oxford University Press.

Charnov, E. L. & Berrigan, D. (1990). Dimensionless numbers and life history evolution: age of maturity versus the adult lifespan. *Evolutionary Ecology* **4**, 273–275.

Chase, M. W. & 41 other authors (1993). Phylogenetics of seed plants: an analysis of nucleotide sequences from the plastid gene *rbc*L. *Annals of the Missouri Botanical Garden* **80**, 528–580.

Chaw, S. M., Sung, H. M., Long, H., Zharkikh, A. & Li, W. H. (1995). The phylogenetic positions of the conifer genera *Amentotaxux*, *Phyllocladus*, and *Nageia* inferred from 18S rRNA sequences. *Journal of Molecular Evolution* **41**, 224–230.

Cochran, M. E. & Ellner, S. (1992). Simple methods for calculating age-based life history parameters for stage-structured populations. *Ecological Monographs* **62**, 345–364.

Coddington, J. A. (1992). Avoiding phylogenetic bias. *Trends in Ecology and Evolution* **7**, 68–69.

Cronquist, A. (1981). *An integrated system of classification of the flowering plants*. New York: Columbia University Press.

Cronquist, A., Takhtajan, A. & Zimmerman, W. (1966). On the higher taxa of Embryobionta. *Taxon* **15**, 129–134.

Dick, J. M., Leakey, R. R. B. & Jarvis, P. G. (1990). Influence of female cones on the vegetative growth of *Pinus contorta* trees. *Tree Physiology* **6**, 151–163.

Enright, N. J., Franco, M. & Silvertown, J. (1995). Comparing plant life histories using elasticity analysis: the importance of lifespan and the number of life-cycle stages. *Oecologia* **104**, 79–84.

Eriksson, O. (1985). Reproduction and clonal growth in *Potentilla anserina* L. (Rosaceae): the relation between growth form and dry weight allocation. *Oecologia* **66**, 378–380.

Evans, F. C. & Smith, F. E. (1952). The intrinsic rate of natural increase for the human louse, *Pediculus humanus* L. *American Naturalist* **86**, 299–310.

Felsenstein, J. (1985). Phylogenies and the comparative method. *American Naturalist* **125**, 1–15.

Fiedler, P. L. (1987). Life history and population dynamics of rare and common mariposa lilies (*Calochortus* Pursh: Liliaceae). *Journal of Ecology* **75**, 977–995.

Fone, A. L. (1989). A comparative demographic study of annual and perennial species of *Hypochoeris* (Asteraceae). *Journal of Ecology* **77**, 495–508.

Franco, M. & Silvertown, J. (1990). Plant demography: what do we know? *Evolutionary Trends in Plants* **4**, 74–76.

Franco, M. & Silvertown, J. (1994). On trade-offs, elasticities and the comparative method: a reply to Shea, Rees & Wood. *Journal of Ecology* **82**, 958.

Frascaria, N., Maggia, L., Michaud, M. & Bousquet, J. (1993). The *rbc*L gene sequence from chestnut indicates a slow rate of evolution in the Fagaceae. *Genome* **36**, 668–671.

Garland, T., Jr. (1992). Rate tests for phenotypic evolution using phylogenetically independent contrasts. *American Naturalist* **140**, 509–519.

Geber, M. A. (1990). The cost of meristem limitation in *Polygonum arenastrum*: negative genetic correlations between fecundity and growth. *Evolution* **44**, 799–819.

Gren, J. & Willson, M. F. (1994). Cost of seed production in the perennial herbs *Geranium maculatum* and *G. sylvaticum* – an experimental field study. *Oikos* **70**, 35–42.

Gunter, L. E., Kochert, G. & Giannasi, D. E. (1994). Phylogenetic relationships of the Juglandaceae. *Plant Systematics and Evolution* **192**, 11–29.

Hamilton, W. D. (1966). The moulding of senescence by natural selection. *Journal of Theoretical Biology* **12**, 12–45.

Harper, J. L. & White, J. (1974). The demography of plants. *Annual Review of Ecology and Systematics* **5**, 419–463.

Harvey, P. H. & Pagel, M. D. (1991). *The comparative method in evolutionary biology*. Oxford University Press.

Harvey, P. H. & Zammuto, R. M. (1985). Patterns of mortality and age at first reproduction in natural populations of mammals. *Nature* **315**, 319–320.

Jordano, P. (1995). Angiosperm fleshy fruits and seed dispersers: a comparative analysis of adaptation and constraints in plant-animal interactions. *American Naturalist* **145**, 163–191.

Kawano, S., Takada, T., Nakayama, S. & Hiratsuka, A. (1987). Demographic differentiation and life history evolution in temperate woodland plants. In *Differentiation patterns in higher plants* (ed. K. M. Urbanska), pp. 153–181. London: Academic Press.

Kelly, C. K. (1995a). Seed size in tropical trees: a comparative study of factors affecting seed size in Peruvian angiosperms. *Oecologia* 102, 377–388.

Kelly, C. K. (1995b). Thoughts on clonal integration: facing the evolutionary context. *Evolutionary Ecology* 9, 575–585.

Kelly, C. K. & Beerling, D. J. (1995). Plant life form, stomatal density and taxonomic relatedness: a reanalysis of Salisbury (1927). *Functional Ecology* 9, 422–431.

Kelly, C. K. & Purvis, A. (1993). Seed size and establishment conditions in tropical trees: on the use of taxonomic relatedness in determining ecological patterns. *Oecologia* 94, 356–360.

Kroon, H. de, Plaiser, A., Groenendael, J. van & Caswell, H. (1986). Elasticity: the relative contribution of demographic parameters to population growth rate. *Ecology* 67, 1427–1431.

Mazer, S. J. (1990). Seed mass of Indiana Dune genera and families: taxonomic and ecological correlates. *Evolutionary Ecology* 4, 326–357.

Mesterton-Gibbons, M. (1993). Why demographic elasticities sum to one: a postscript to de Kroon *et al. Ecology* 74, 2467–2468.

Olmstead, R. G. & Reeves, P. A. (1995). Evidence for the polyphyly of the Scrophulariaceae based on chloroplast *rbc*L and *ndh*F sequences. *Annals of the Missouri Botanical Garden* 82, 176–193.

Piñero, D., Sarukhán, J. & Alberdi, P. (1982). The costs of reproduction in a tropical palm, *Astrocaryum mexicanum. Journal of Ecology* 70, 473–481.

Promislow, D. E. L. & Harvey, P. H. (1990). Living fast and dying young: a comparative analysis of life history variation among mammals. *Journal of Zoology* 220, 417–437.

Purvis, A. & Rambaut, A. (1995). Comparative analysis by independent contrasts (CAIC): an Apple Macintosh application for analysing comparative data. *Computer Applications in the Biosciences* 11, 247–251.

Pyke, G. H. (1991). What does it cost a plant to produce floral nectar? *Nature* 350, 58–59.

Read, A. F. & Harvey, P. H. (1989). Life history differences among the eutherian radiations. *Journal of Zoology* 219, 329–353.

Renner, S. S. & Ricklefs, R. E. (1995). Dioecy and its correlates in the flowering plants. *American Journal of Botany* 82, 596–606.

Roff, D. A. (1992). *The evolution of life histories.* London: Chapman & Hall.

Rose, M. R. (1991). *Evolutionary biology of aging.* Oxford University Press.

Ross, C. (1992). Environmental correlates of the intrinsic rate of natural increase in primates. *Oecologia* 90, 383–390.

Sackville-Hamilton, N. R., Schmid, B. & Harper, J. L. (1987). Life history concepts and the population biology of clonal organisms. *Proceedings of the Royal Society of London* **B 232**, 35–57.

Sarukhán, J. & Gadgil, M. (1974). Studies on plant demography: *Ranunculus repens* L. *R. bulbosus* L., and *R. acris* L. III. A mathematical model incorporating multiple modes of reproduction. *Journal of Ecology* **62**, 921–936.

Shea, K., Rees, M. & Wood, S. N. (1994). Trade-offs, elasticities and the comparative method. *Journal of Ecology* **82**, 951–957.

Silvertown, J. & Franco, M. (1993). Plant demography and habitat: a comparative approach. *Plant Species Biology* **8**, 67–73.

Silvertown, J., Franco, M. & McConway, K. (1992). A demographic interpretation of Grime's triangle. *Functional Ecology* **6**, 130–136.

Silvertown, J., Franco, M. & Menges, E. (1996). Interpretation of elasticity matrices as an aid to the management of plant populations for conservation. *Conservation Biology* **10**, 591–597.

Silvertown, J., Franco, M. & Pérez-Ishiwara, R. (1997). The evolution of senescence in perennial plants. (Submitted)

Silvertown, J., Franco, M., Pisanty, I. & Mendoza, A. (1993). Comparative plant demography – relative importance of life-cycle components to the finite rate of increase in woody and herbaceous perennials. *Journal of Ecology* **81**, 465–476.

Southwood, T. R. E. (1988). Tactics, strategies and templets. *Oikos* **52**, 3–18.

Stearns, S. C. (1992). *The evolution of life histories*. Oxford University Press.

Sutherland, W. J., Grafen, A. & Harvey, P. H. (1986). Life history correlations and demography. *Nature* **320**, 88.

Svensson, B. M., Carlsson, B. A., Karlsson, P. S. & Nordell, K. O. (1993). Comparative long-term demography of three species of *Pinguicula*. *Journal of Ecology* **81**, 635–645.

Tienderen, P. H., van. (1995). Life cycle trade-offs in matrix population models. *Ecology* **76**, 2482–2489.

Watkinson, A. R. & White, J. (1985). Some life-history consequences of modular construction in plants. *Philosophical Transactions of the Royal Society of London* **B 313**, 31–51.

Woodland, D. W. (1991). *Contemporary plant systematics*. Englewood Cliffs, NJ: Prentice Hall.

12 · Life history evolution in heterogeneous environments: a review of theory

Richard M. Sibly

1 Introduction

The world is variable in space and time, and this inevitably affects the evolutionary process. Quite how is harder to say. Theoreticians have addressed the question using the tools of population genetics, population dynamics and, more recently, life history theory, as follows.

Population genetic models take full account of differences in performance of different genotypes, and a primary interest is in how level of dominance affects the maintenance of genetic polymorphism. The two principal schools modelling spatial heterogeneity differ as to whether or not they allow density dependence within habitats. The two extreme cases are referred to as 'soft' and 'hard' selection respectively. Soft selection models date from a paper by Levene (1953) (see also Wallace 1968), and hard selection models from Dempster (1955). Temporal variation has also been considered following the discovery by Haldane & Jayakar (1963) that differences in genotype performance during occasional catastrophes can also promote the persistence of polymorphisms. Recent reviews can be found in Hartl & Clark (1989) and Barton & Clark (1990). The life history is implicity assumed to be semelparous in these models and the generations are assumed not to overlap.

Levins (1962, 1963, 1968) introduced 'fitness sets' as an alternative way to think about the evolutionary effects of heterogeneity without worrying about levels of dominance, and evolutionary ecologists have long found this a useful approach. The relationship with population genetics is, however, not always clear (Endler, 1977; Ricklefs, 1990). One topic of special interest has been Levins' analysis of the effects of varying the frequency of two habitats, and his demonstration that a gradual cline of habitat frequency can produce an abrupt change in phenotype.

The study of life history evolution owes much to three influential papers (Charnov & Shaffer 1973; Schaffer 1974*a, b*) which set the scene for much

228

subsequent analysis of the evolutionary implications of trade-offs. In one of these Schaffer (1974b) considered the impact of temporal environmental variation on the optimal life history, and the strategy of spreading one's investment between different broods in a variable environment has since become known as bet-hedging (Stearns 1976). Life history analysis of evolution in spatially heterogeneous environments has been revolutionized by the discovery by Kawecki & Stearns (1993) and Houston & McNamara (1992) that earlier treatments had used incorrect fitness measures. This has important implications for the analysis of phenotypic plasticity (the ability of individuals in heterogeneous environments to modify their phenotypes according to the kind of habitat they find themselves in). In particular, correct identification of the optimal 'reaction norm' depends on use of an appropriate fitness measure. The quantitative genetics of the evolution of reaction norms has been described by Via & Lande (1985) and de Jong (1990).

Identification of appropriate fitness measures has also been a focus of research on temporally varying environments. Using matrix representation of the life history, environmentally induced variations in the life history can be represented by sequences of matrices. Diploidy is generally ignored, and the prime interest is in what happens to the population in the long run, and in how to assess this in terms of the population's growth rate and its probability of extinction. Little has been achieved in incorporating the effects of density dependence. These matrix analyses were initiated by Cohen (1976) and Tuljapurkar (1982), and are reviewed in Caswell (1989).

Here I review what is known of life history evolution constrained by trade-offs, in heterogeneous environments. After a brief outline of the life history process, spatial heterogeneity is considered, then temporal variation. The variation between habitats may be of two kinds. The shape or position of the trade-off may vary between habitats. Alternatively the trade-off may be fixed but other life history traits may vary. The applicability of existing results to plants may be limited by the assumptions made in the analysis. This problem and its possible resolutions are discussed in the final section.

2 Life history evolution

Life history evolution operates through the selection of alleles coding for some trait or combination of traits, at the expense of competitor alleles coding for other combinations of traits. What is meant when an allele is said to be 'selected' is that its numbers increase at the expense of those of competitor alleles, if there are competitor alleles, and if there are not – in which case the allele is at fixation – then the allele resists invasion by such competitor alleles as may arise through mutation. The key question is whether the numbers of each allele increase or decrease in the whole population. The rates of increase or decrease of alleles therefore have to be studied, and these are most conveniently calculated on a 'per copy' basis. Thus the 'per copy rate of increase' of an allele in a specified environment is one of its key attributes, and this will be referred to here as the *fitness* of the allele, following Sibly & Curnow (1993). Although this definition of fitness is closely related to that used in classical population genetics, the two are not identical (for further discussion see Sibly & Curnow 1993).

The fitness, or per copy rate of increase in the numbers, of an allele can also be thought of as the population growth rate of the allele. There is a good analogy between the study of populations of individuals and that of populations of alleles, and the analytic techniques appropriate to the study of one are often also applicable to the other. In particular this is true in calculating population growth rate from knowledge of individual life histories. In population dynamics, population growth rate, sometimes referred to as the 'Malthusian parameter', here designated F, is related to the life histories of the individuals in the population by the Euler-Lotka equation, as follows. Suppose the individuals reproduce at ages $t_1, t_2, t_3, \ldots$, then producing $n_1, n_2, n_3, \ldots$, offspring, and suppose that offspring survive from birth to ages $t_1, t_2, t_3, \ldots$, with probabilities $l_1, l_2, l_3, \ldots$ respectively. Then the Euler-Lotka equation is:

$$1 = \sum_i e^{-Ft_i} l_i n_i \tag{1}$$

This equation is used to calculate F from knowledge of the t_is, l_is and n_is. The equation is derived on the supposition that population growth rate is constant during the lifetime of the individuals, or alternatively on the supposition that during individual lifetimes the proportion of individuals that are in each age class is constant. The l_is and n_is can be average values provided that the averages are calculated over all those individuals alive at age 0. However it is assumed that there is no variation in the t_is.

With modified interpretation, the Euler–Lotka equation can also be used to calculate the population growth rate (i.e. fitness) of an allele, where now the life history parameters (the l_is and n_is) are the average values of the carriers of the allele (the averages being computed over all copies of the allele present in individuals alive at age 0). A formal derivation and proof is provided by Sibly & Curnow (1993).

During the evolutionary process, alleles increase in numbers in populations if they have higher fitness than their competitor alleles, and once at fixation this ability implies the ability to resist invasion by competitors. The outcome of the evolutionary process is, therefore, that there remain in the population only those alleles with highest fitness. The life histories of the carriers of these alleles are termed *optimal life histories* because they confer maximum fitness in the study environment, and the alleles can there be thought of as *optimal alleles*. Optimal life histories are, therefore, expected outcomes of the evolutionary process.

The main limitation of the above approach is the assumption that the average life history of the carriers of an allele remains constant over time. This may be approximately true in a tightly regulated population in which the optimal alleles have reached fixation, so that their fitness is zero and remains at zero. However, it cannot be exactly true of an allele invading a population as then its fitness is positive, but when it approaches fixation, its fitness must decline to zero. Thus fitnesses may change when new alleles invade populations, and this will be associated with changes in individual life histories. Population growth rates, and thus fitnesses, may also change as a result of density-dependent processes.

This in outline summarizes the operation of the life history evolutionary process in a uniform environment. The central concept is the fitness of an allele, measured by its population growth rate, and related to the life history of its carriers by equation (1). In the next section we consider how the process is modified if the population lives in a spatially-heterogeneous environment.

2 Spatially heterogeneous environments

An attractive starting point is the case of a spatially heterogeneous environment consisting of a number of habitats in which the life histories of genetically identical individuals differ. Let the habitats be labelled $0, 1, 2, 3, \dots, H$. Consider the simple case that:

Assumption 1: gametes disperse randomly between habitats.

Eventually some gametes pair up to form zygotes. Thereafter:

Assumption 2: zygotes stay in their initial habitat until they die.

If a proportion $q(h)$ of newly-formed zygotes are in habitat h, and if $b_i(h)$ represents the product of l_i and n_i in habitat h, then the Euler–Lotka equation, equation (1), takes the form:

$$1 = \sum_{h,i} q(h) \, e^{-Ft_i} b_i(h) \tag{2}$$

(Kawecki & Stearns, 1993; Sibly, 1995). This equation can be used to calculate the fitness of each allele and to identify optimal alleles/life histories, just as was done above in the case of uniform environments.

It may happen that the optimal life histories change if one or more habitats are isolated from the others. In this case an isolated habitat optimum (*isolated optimum*) can be identified, which may be distinct from the global optimum. When the isolated and global optima differ, the position of the global optimum depends on the relative frequencies of the different habitat types – the $q(h)$s in equation (2). The form of the dependency has been addressed in some stimulating papers by Levins (1963, 1963, 1968), but as mentioned earlier no satisfactory method exists by which to include Levins' approach within the framework of conventional evolutionary genetics. In a recent paper, however, Sibly (1995) tackled some simple cases numerically, as described below.

To gain some initial insight into the operation of selection in heterogeneous environments Sibly (1995) analysed a simple five-parameter life cycle characterized by juvenile and adult survivorships, age at maturity, interval between breeding attempts and fecundity, which was assumed to be the same at each breeding attempt. The model can also be formulated in terms of mortality rates rather than survivorships, if appropriate. Adult size is implicitly assumed constant because fecundity is the same at each breeding attempt, and adult mortality rate is assumed constant.

Different trade-offs were assumed to operate in different habitats. To keep things simple, only two habitat types were considered, labelled 0 and 1. The trade-offs operating in each habitat type are shown in Figure 12.1. The forms of these trade-offs were chosen so that whilst broadly plausible (Sibly & Calow 1986; Sibly & Antonovics 1992) they give clear fitness landscapes (Figure 12.2) with the isolated optima being clearly distinct and each having fitness ≈ 0.

The fitness landscapes that result from the trade-offs of Figure 12.1 are shown in Figure 12.2. For each trade-off a three-dimensional representation of the fitness landscape is shown on the left and a two-dimensional fitness-contour map on the right. In each case the z-axis (the 'vertical' axis) is fitness and the x-axis represents the frequency of habitat 1, i.e. the proportion of patches that are of the type of habitat 1. The y-axis represents a trait affecting the life history, and this varies according to the trade-off being considered. In constructing the landscapes shown in Figure 12.2 it was assumed that individuals are not phenotypically plastic, i.e. do not modify their phenotypes according to the kind of habitat they find themselves in. Following Sibly (1995) we refer to a trait of this type as being 'aplastic'.

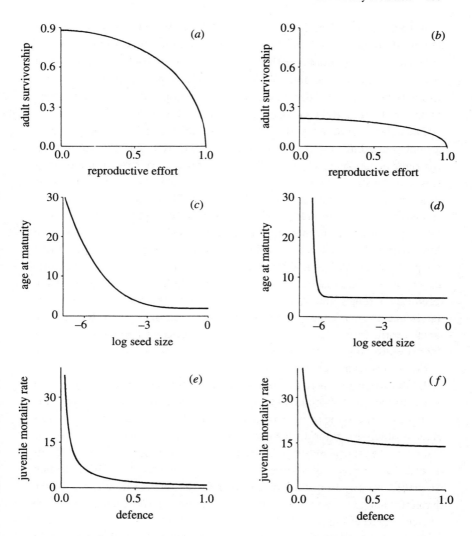

Figure 12.1. The forms of three possible trade-offs in each of two habitats. The cost of reproduction trade-off in habitat 0 (a) and habitat 1 (b), in which as reproductive effort increases, adult survivorship decreases. Note that the form of the relationship differs between the two habitats. A seed-size trade-off is shown in habitat 0 (c) and in habitat 1 (d). Because seed size is plotted on a log scale, −6 refers to very small seeds and 0 to relatively large seeds. It is assumed that it takes longer for individuals from smaller seeds to reach maturity, hence the negative slope of the trade-off. A trade-off between mortality rate and expenditure on defence is shown in habitat 0 (e) and in habitat 1 (f). Increasing expenditure on defence results in a reduction in mortality rate, giving the negative slope of the trade-off. After Sibly (1995).

If the frequency of habitat 1 is 0, then the environment consists entirely of habitat 0, so habitat 0 is effectively isolated, and the fitness optimum is the isolated fitness optimum. In the reproductive effort trade-off (top row of Figure 12.2) the isolated optimum in habitat 0 is a reproductive effort of 0.50. Conversely, the isolated optimum in habitat 1 is observed when the frequency of habitat 1 is 1, and this occurs at a reproductive effort close to 1.

The aplastic optima (i.e. the optimal aplastic alleles) for spatially hetero-geneous environments are shown by the dotted lines on the contour plots on the right-hand side of Figure 12.2. In the reproductive effort trade-off (top row) the aplastic optima switch gradually from one isolated optimum to the other, as the frequency of habitat 1 increases. However, in the other two trade-offs the optimal aplastic strategy switches abruptly between the iso-lated optima. The abruptness or otherwise of the switch appears to depend on how sharply the fitness peaks are differentiated. In the top row of Figure 12.2 the peaks in the isolated habitats are relatively rounded, whereas in the other cases the peaks are sharp and clearly differentiated. This sharp differentiation produces a ridges and valley system parallel to the x-axis in the fitness landscape; the existence of the valley between the ridges makes the optimal aplastic strategy switch abruptly between the isolated optima as habitat 1 increases. By contrast, the top row of Figure 12.2 lacks a valley parallel to the x-axis, so the transition is gradual.

In all three cases, then, the optimal aplastic strategy is between or almost between the isolated optima, and the global optimum switches between the isolated optima as the relative frequencies of the habitats change. The abrupt-ness or otherwise of the switch depends on the position and structure of the valleys in the fitness landscape. Sharp differentiation of the fitness peaks leads to deep valleys between the fitness ridges, and in this case the global optimum switches abruptly from one to the other as the relative frequencies of the habitats change. However, if fitness peaks are not sharply differentiated, the switch may be gradual.

It would be attractive if it were possible to derive the fitness landscape from knowledge of the shape of the fitness peaks in isolated habitats. Some progress can be made by considering the following simple model lacking age structure. The argument will be phrased in terms of individuals, but 'allele copies' can be substituted for individuals to obtain an allelic version if desired. Suppose there are $N_h(t)$ individuals (or allele copies) in habitat h at time t, and suppose that this population would increase by δN_h between t and $t + \delta t$ if habitat h was isolated. δN_h is the difference between the number of new individuals produced in habitat h, δB_h, say, and the number that die, δM_h.

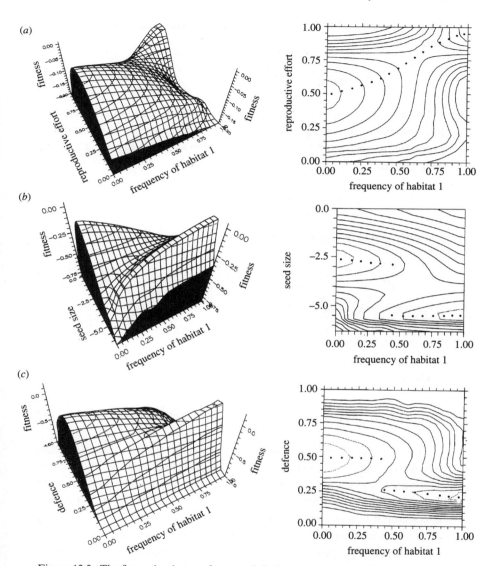

Figure 12.2. The fitness landscapes for a spatially heterogeneous environment consisting of two habitat types specified by the trade-offs shown in Figure 12.1. Each row refers to a different trade-off as in Figure 12.1. (*a*) The trade-off involving reproductive effort; (*b*) the seed-size trade-off; (*c*) the defence trade-off. Dotted lines on the contour plots indicate optimal aplastic strategies. Optimal isolated strategies occur when the frequency of habitat 1 is 0 or 1. The graphical procedure (using UNIMAP) involved smoothing and this occasionally produced minor misrepresentations. See text for further discussion. After Sibly (1995).

Thus $\delta N_h = \delta B_h - \delta M_h$. Let F_h be the fitness of individuals in habitat h if it were isolated. Thus:

$$\frac{dN_h}{dt} = F_h N_h(t) \tag{3}$$

Consider now what would happen if the habitats were not isolated. The total population at time t is still $\Sigma_h N_h(t)$. The total number of new individuals produced between t and $t + \delta t$ is $\Sigma_h \delta B_h$ and the total number of deaths is $\Sigma_h \delta M_h$. Hence:

$$\delta N = \sum_h \delta B_h - \sum_h \delta M_h = \sum_h \delta N_h \tag{4}$$

and since global fitness, F, is defined by:

$$\frac{dN}{dt} = FN \tag{5}$$

it follows from equations (3) to (5) that

$$\delta N = FN\delta t = \sum_h \delta N_h = \sum_h F_h N_h \delta t$$

Hence:

$$F = \sum_h F_h \frac{N}{N} \tag{6}$$

This shows that F is simply the weighted average of the F_hs, each F_h being weighted by N_h/N, i.e. by the proportion of individuals that live in habitat h. This may provide a good approximation to the age-structured model when the isolated-habitat fitnesses are similar (Sibly 1995). However, when isolated-habitat fitnesses differ markedly, ignoring age structure can give very different results from the age-structured model, as will now be shown.

In comparing the models with and without age structure, note first that the proportion of individuals that live in habitat h, N_h/N (equation (6)), is not the same as q_h (equation (2)), which is defined as the proportion of newly formed zygotes that occur in habitat h. The two quantities are the same only if the death rates in all habitats are the same. Nevertheless the two are obviously related, and, like q_h, N_h/N can be regarded as a measure of the 'frequency of habitat 1'. It is then of interest to plot the analogue of Figure 12.2 for the model lacking age structure, with the x-axis ('frequency of habitat 1') repre-

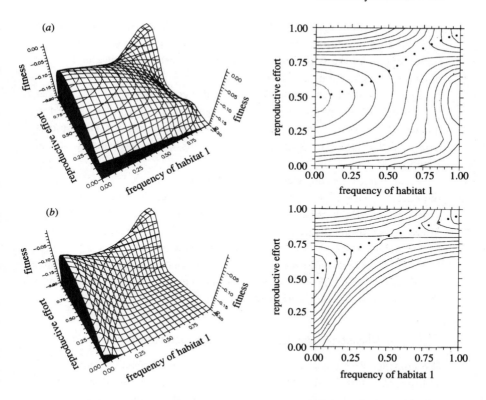

Figure 12.3. Fitness landscapes obtained from spatially heterogeneous models (*a*) with and (*b*) without age structure, for the trade-offs shown in Figures 12.1*a* and *b* and 12.2*a*. Dotted lines on the contour plots represent the optimal aplastic strategies. The landscape in the bottom row also has an interpretation for age-structured populations in temporally heterogeneous environments (see equation (8)).

senting N_h/N instead of q. A comparison is made in Figures 12.3 and 12.4 between the model with age structure (and $x = q_h$, top rows of Figures 12.3 and 12.4) and the model without age structure (and $x = N_h/N$, bottom rows of Figures 12.3 and 12.4).

Figure 12.3 shows the reproductive effort model of figure 1*a* and *b*. Note that dependence of fitness on reproductive effort in the isolated habitats (i.e. at frequencies of habitat 1 of 0 or 1) is the same whether age structure is included (top row) or not (bottom row). Taking a cross-section through the landscape parallel to the *x*-axis results in a straight line in the model without age structure (bottom row), and this is the result of the linear dependence of fitness on N_h/N in equation (6). By contrast a cross-section through the age-structured landscape (top row) is far from a straight line. Despite this, the way the aplastic optima change as the frequency of habitat 1 increases is rather similar in the two models (compare the dotted lines in the contour plots on the right-hand side of Figure 12.3).

Figure 12.4 shows that qualitatively similar conclusions hold for the defence trade-off, except that the way the aplastic optima change as the frequency of habitat 1 increases differs drastically between the two models (compare the dotted lines in the contour plots on the right-hand side of Figure 12.4).

In summary, it seems essential to include age structure when calculating the effects of selection in spatially-heterogeneous environments, when the fitness peaks are sharply differentiated as in Figure 12.4. However if the fitness peaks are relatively rounded, as in Figure 12.3, the two models give fairly similar results, at least in computing aplastic optima.

So far we have not allowed the possibility of phenotypic plasticity, which describes the ability of phenotypes to adjust their life histories according to the habitat in which they find themselves.

4 Phenotypic plasticity in spatially structured environments

We consider first the case of 'complete' or 'infinite' plasticiy (Sibly 1995; Houston & McNamara 1992) meaning that strategy in any one habitat is completely independent of strategy in any other. With complete plasticity the expected evolutionary outcome is an optimal reaction norm, a concept discussed by Stearns & Koella (1986). Complete plasticity cannot evolve if habitats are isolated and identification of the optimal reaction norm must therefore take account of the flow of alleles between habitats (Houston & McNamara 1992; Kawecki & Stearns 1993). Earlier analyses did not take this into account and as a results their analyses of optimal reaction norms were flawed.

Kawecki & Stearns (1993) introduced the important theorem that the optimal plastic strategy maximizes the reproductive value of newborns in each habitat (a proof is provided in the Appendix). This theorem is most easily understood in the case that the global population is stable and all alleles are at evolutionary equilibrium, so that $F = 0$. Reproductive value at birth is then the same as lifetime reproductive success (LRS), that is, the expected number of offspring that a newborn will produce over its lifetime. In the case that $F = 0$, the theorem states that the optimal plastic strategy is to produce as many offspring as possible in each habitat. The theorem is

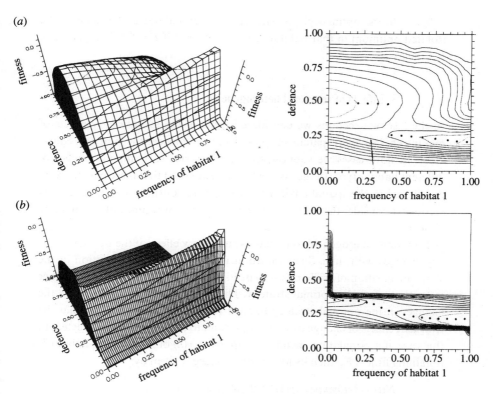

Figure 12.4. Fitness landscapes obtained from models (a) with and (b) without age structure, for the trade-offs shown in Figures 12.1e and f and 12.2c. Other details as in Figure 12.3.

intuitive when one remembers the assumptions that gametes disperse randomly between habitats, and that newborn stay in their natal habitats. Nevertheless, the validity of the theorem was not generally appreciated until it was pointed out by Kawecki & Stearns (1993).

When $F \neq 0$ the reproductive value of newborns corresponds to discounted LRS, in which case offspring born at age t_i are discounted by e^{-Ft_i}. Thus in growing populations $(F > 0)$ later offspring are discounted, and the faster the population grows, the higher the discount. This is because later offspring have less genetic value, because they are inserted into an expanded gene pool. Their contribution to parental reproductive value is therefore lower. Conversely, in declining populations $(F < 0)$ later offspring are at a premium and contribute more to parental reproductive value.

If $F = 0$, the optimal plastic strategy in each habitat is the same as the isolated optimum in that habitat (Sibly, 1995), but if $F \neq 0$ this is generally not the case.

5 Effects of temporal heterogeneity

In the last two sections we considered some of the evolutionary effects of making environments patchy in space. Here we consider the temporal equivalent. In both cases we suppose that life histories vary among patches of different habitat type. In the last section we supposed that these patches were distributed in the spatial environment but were constant over time. In this section we consider the converse, that patches vary over time but are spatially uniform.

Temporal heterogeneity only has evolutionary effects if the optimal strategy changes with time. Such variation may be a result of variation in the shape or position of trade-offs – as in Figure 12.1 – or of variation in other traits that affect the optimal strategy. The simplest case is that in which the changes in the optimal strategy happen relatively infrequently, so that the transient effects of change from one optimum to another can be neglected. In this case, if a population spends a proportion $p(h)$ of its time t in habitat h, obtaining fitness F_h, then its long term increase is given by:

$$N(t) = N(0)\exp(F_1 p(1)t + F_2 p(2)t + \dots)$$

$$= N(0)\exp(\Sigma_h F_h p(h)t) \tag{7}$$

so its long-term rate of increase, $\bar{F}$, is:

$$\bar{F} = \frac{\sum_h F_h p(h)t}{t} = \sum_h F_h p(h) \tag{8}$$

which is of the form of equation (6), with N_h/N replaced by $p(h)$. Hence $\bar{F}$ is simply the weighted average of the F_hs, each F_h being weighted by $p(h)$, where $p(h)$ represents the proportion of time that the population spends living in habitat h. The relationship is illustrated for the case of two habitats by the bottom rows of Figures 12.3 and 12.4, where now the x-axis indicates the proportion of its time that the population spends in habitat 1. Note that for any given value of x (e.g. for any given sequence of habitats) there is only one optimum. In these examples fitness falls off sharply on either side of the optimum at all values of x (except at $x = 0$ in Figure 12.4). Thus there is never any indication in these cases that temporal heterogeneity might allow polymorphisms to persist.

Another obvious case to consider is that of random fluctuations in life history traits. The analytic theory uses matrix representation of populations and has been reviewed by Tuljapurkar (1990), Caswell (1989), and Metz *et al.* (1992). The sequence of habitats encountered is represented by a sequence of matrices, one for each time unit. The general conclusion is that, under certain assumptions, the long-term rate of increase of the population is best measured by a generalization of F known as the dominant Lyapunov exponent. In the case of small independent fluctuations in life history traits this can be expanded about its value (F_c, say) in a constant environment. Remember that the sequence of habitats encountered is represented by a sequence of matrices, one for each time unit. For each matrix a value of F can be calculated, let the variance of these numbers be written $V(F)$. Then it turns out that the Lyapunov exponent is approximately equal to

$$F_c - \frac{V(F)}{2\exp(F_c)}$$

(Caswell, 1989, p. 224). Autocorrelation between successive environments can be dealt with similarly provided that environmental fluctuations are small. However, for some life histories small fluctuations in life history traits have no effect on long-term fitness, so optimal strategies/evolutionary outcomes are unaffected (Sibly *et al.* 1991).

The evolutionary effects of large fluctuations in life history traits have not been widely examined, except in the case of the reproductive effort trade-off. Even here it has been assumed that the trade-off curve itself is not affected by the fluctuations. The original analysis was due to Schaffer (1974*b*). Using a simple analytical model Schaffer showed that optimal reproductive effort should decrease in an environment in which juvenile survivorship varied from year to year. In other words, in a variable world not too many eggs should be placed in one 'basket' (year). This strategy was termed 'bet-hedging' by Stearns (1976). However, if adult survivorship varies, then optimal reproductive effort should increase. Thus environmental variation leads to an increase or decrease in optimal reproductive effort according to whether it is adult or juvenile survivorship which is variable.

Bet-hedging has often been invoked by field ecologists as a possible factor affecting the interpretation of their results (see for example Godfray *et al.* 1991). However, its true significance has been hard to evaluate, partly because it is difficult to obtain the necessary measurements of survivorship variances and of the shape of the trade-off curve, and partly because of restrictive assumptions made in the original analysis.

However, Sibly *et al.* (1991) and Cooch & Ricklefs (1994), using computer simulations, have shown that Schaffer's conclusions are valid beyond the range of the restrictive assumptions used in the original analysis. Sibly *et al.* (1991) showed for sexually reproducing diploid populations that whether random variation occurred in juvenile or adult survivorship from either a normal or uniform distribution, there was relatively little effect on optimal reproductive effort, even for large variations with standard deviation equal to the mean (Figure 12.5, top row). Where optimal reproductive effort was affected, it moved as predicted by Schaffer (1974*b*), particularly when asymmetries in variation were accounted for (Figure 12.5). Perhaps surprisingly, temporal heterogeneity had little effect on the speed of convergence to the eventual evolutionary outcome (Sibly *et al.* 1991).

Cooch & Ricklefs (1994) showed similarly that large random variations in fecundity had relatively little effect on the evolution of reproductive effort. In addition they showed that density dependence and temporal environmental autocorrelation acted to reduce the effects of variation. They also found that the shape of the trade-off curve did not affect the conclusions, provided the trade-off curve was sufficiently convex.

These results may, however, require modification if the population dynamics are non-linear, for instance chaotic (Ferriere & Fox 1995).

6 Discussion

The results pertaining to temporal heterogeneity are the most straightforward and will be discussed first. It seems that if the trade-off is fixed, then even large stochastic variations in life history traits have little effect on evolutionary outcomes or rates. Such effects as do occur are well predicted by Schaffer's (1974*b*) model, with modification for asymmetric variations as necessary. Unfortunately, the period of study needed to observe such effects is so long as to preclude experimental testing for most organisms. Reviews of such animal evidence as is available can be found in Roff (1992) and Stearns (1992).

To date no analysis has been made of the effects of temporal variation in the trade-off. A problem in designing such an analysis is that it is difficult to know what sorts of variation are plausible. Data on this point are hard to obtain because it is difficult to establish experimentally shapes of trade-offs even in constant environments (with the possible exception of resource allocation trade-offs; see papers in Bazzaz & Grace 1997).

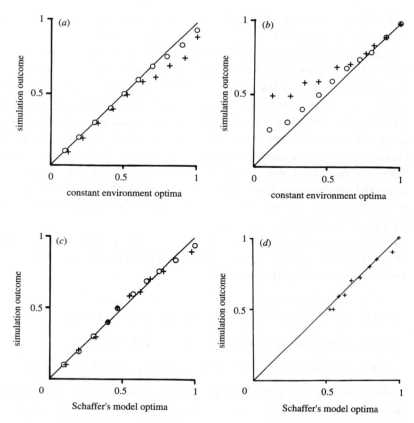

Figure 12.5. Effects of temporal variability in juvenile (a) and (c) or adult (b) and (d) survivorship on optimal reproductive effort. A constant trade-off of the general form of figure 1a was assumed. Outcomes after 5000 generations of simulated evolution in a variable environment are plotted against predictions from a modified version of Schaffer's analytical model. Variations were chosen randomly from normal distributions, with the standard deviation of the variations set equal to their mean (denoted by crosses) or half the mean (denoted by open circles). Choosing variations from uniform distributions gave very similar results. From Sibly et al. (1991).

The results on spatial heterogeneity have many ramifications. The three-dimensional method presented here (Figure 12.2) supersedes Levins' (1962, 1963, 1968) two-dimensional fitness set analysis, as it is based on a rigorous definition of allelic fitness. Contrary to the general understanding of Levins' position (e.g. Rickleffs 1990), fitness in a spatially heterogeneous environment is not simply the average of the fitnesses in individual habitats, at least for age-structured populations. This can be seen by comparing Figures 12.2, 12.3 and 12.4. Nevertheless Levins' interesting prediction that a gradual cline in habitat frequency can produce an abrupt change in phenotype is preserved (Figures 12.2b and c).

Hopefully the present more detailed theoretical treatment of the phenomenon will suggest new ways of testing this prediction. The starting point might be a series of reciprocal transplant experiments measuring the life histories and fitnesses of different strains/genotypes in each of a number of habitats (e.g. Stratton 1994). The ways in which allele frequencies actually change along environmental gradients and the mechanisms responsible for those changes have been reviewed in depth by Endler (1977). Extension of the theoretical results of Figure 12.2 to the cases of sources and sinks, transition zones and density-dependence are briefly considered in Sibly (1995) (see also Holt 1985; McGinley 1987; Pulliam 1988; Pulliam & Danielson 1991; van Tienderen 1991; Holt & Gaines 1992).

A worrying feature of the analysis of spatial heterogeneity is its dependence on assumptions that are implausible for many plants. In particular, two assumptions have already been noted:

Assumption 1: gametes disperse randomly between habitats;
Assumption 2: zygotes stay in their initial habitat until they die.

In the analysis underlying Figure 12.2 it was assumed that:

Assumption 3: within each habitat the life history is simple and regular.

The possibility that within-individual variation might be adaptive is neglected, even though various authors have suggested that within-individual variation in seed size may have evolved as a bet-hedging strategy in a variable environment (see for example Geritz 1995). Of the above assumptions, only Assumption 2 is generally plausible and does not need further discussion. The implications of Assumption 3 will be discussed before those of Assumption 1.

Plant life histories are, with some exceptions, generally not simple and regular (Assumption 3). Although the present results may provide a useful starting point, it is obviously necessary to explore more realistic model life histories. The problem here is to identify appropriate models. Perhaps breeding should occur at regular intervals as above, but fecundity should increase exponentially with age, reflecting an increase in somatic size.

For Assumption 1 to be plausible, it would be necessary for dispersal distances to be much greater than patch 'diameters'. Otherwise many gametes would not disperse outside the parental patch. It is known that seed dispersal is in many species of the order of metres or more (Howe & Westley 1986), and some recent work suggests that patch diameters may be less than this. For instance, Stratton (1994), studying an early successional weed *Erigeron annuus*, showed that almost all G × E interactions for fitness occurred at the smallest spatial scale studied (10 cm). According to the information given, over half the seeds dispersed more than 1.5 m. In consequence progeny were likely to disperse to other patches (strictly, to microsites where their relative fitness was unpredictable). Similarly Lechowicz & Bell (1991) and Bell *et al.* (1991), studying understorey herbs in an old-growth forest, demonstrated environmental variance at scales (2 m) relevant to seed dispersal (cf. Mitchell-Olds 1992).

Even if Assumption 1 is untrue, however, the fitness landscapes of Figure 12.2 give some insight into the evolutionary process. For any given frequency of habitat 1, x say, more habitat 0 alleles end up in habitat 0 than expected on a chance basis. This means that for alleles doing well in habitat 0, the effective frequency of habitat 1 is less than x. The performance of habitat 0 alleles can then be judged by moving along the habitat 0 ridge from x towards the habitat 0 optimum. A converse shift is needed for habitat 1 alleles. Notice that this shows how traits restricting gamete dispersal are selected for in heterogeneous environments (see also De Meeûs *et al.* 1993). It is a small step to see in this process one of the mechanisms that may lead to sympatric speciation (cf. Brown & Pavlovic 1992).

The fitness effects of limited dispersal also affect the form of the optimal reaction norm (the evolutionary outcome as regards phenotypic plasticity). With random dispersal the optimal within-habitat strategy maximises Σ_i $e^{-Ft_i}b_i$, where F is global fitness. With no dispersal the optimal within-habitat strategy maximises Σ_i $e^{-F_h t_i}b_i$, where F_h is within-habitat fitness. These strategies differ in the discounts given to later reproduction. With random dispersal, the discount is e^{-Ft_i}, with no dispersal, $e^{-F_h t_i}$. This suggests that with limited dispersal a discount $e^{-F_d t_i}$ might be applied, for some value F_d intermediate between F and F_h. This further suggests that, under plausible assumptions, the optimal reaction norm with limited dispersal would be intermediate between those for the other two cases. Further discussion of the effects of limited dispersal can be found in Sibly (1995).

Also relevant to the theory of optimal reaction norms are the effects of incomplete plasticity. Existing theory handles the cases of aplasticity and complete plasticity, but incomplete plasticity has not been considered. The difficulty is in knowing what kinds of incomplete plasticity might be relevant.

Many of the ambiguities about assumptions can be resolved by reciprocal transplant and common garden experiments. These give information about the extent of plasticity, and about its variation with genotype (G × E interaction) if any. In addition, as noted above, reciprocal transplant experiments can give information about genetic options and associated fitnesses in each environment. It might even be possible to design common-garden experiments that would produce experimentally fitness landscapes like those shown in Figure 12.2.

References

Barton, N. & Clark, A. (1990). Population structure and processes in evolution. In *Population biology: ecological and evolutionary viewpoints* (ed. K. Wöhrmann and S. K. Jain), pp. 115–73. Berlin: Springer-Verlag.

Bazzaz, F. A. & Grace, J.(eds) (1997). *Allocation in plants*. Academic Press. (In the press.)

Bell, G., Lechowicz, M. J. & Schoen, D. J. (1991). The ecology and genetics of fitness in forest plants. III. Environmental variance in natural populations of *Impatiens pallida*. *Journal of Ecology* **79**, 697–713.

Brown, J. S. & Pavlovic, N. B. (1992). Evolution in heterogeneous environments – effects of migration on habitat specialization. *Evolutionary Ecology*, **6**, 360–382.

Caswell, H. (1989). *Matrix population models*. Sunderland: MA: Sinauer.

Charnov, E. L. & Schaffer, W. M. (1973). Life history consequences of natural selection: Cole's result revisited. *American Naturalist* **107**, 791–793.

Cohen, J. E. (1976). Ergodicity of age structure in populations with Markovian vital rates, I: countable states. *Journal of the American Statistical Association*, **71**, 335–339.

Cooch, E. G. & Ricklefs, R. E. (1994). Do variable environments significantly influence optimal reproductive effort in birds? *Oikos*, **69**, 447–459.

De Meeûs, T., Michalakis, Y., Renaud, F. & Olivieri, I. (1993). Polymorphism in heterogeneous environments, evolution of habitat selection and sympatric speciation: soft and hard-selection models. *Evolutionary Ecology*, **7**, 175–198.

Dempster, E. R. (1955). Maintenance of genetic heterogeneity. *Cold Spring Harbor Symposium Quant. Biology* **20**, 25–32.

Endler, J. A. (1977). *Geographic variation, speciation, and clines*. Princeton University Press.

Ferriere, R. & Fox, G. A. (1995). Chaos and evolution. *Trends in Ecology and Evolution*, **10**, 480–485.

Geritz, S. A. H. (1995). Evolutionarily stable seed polymorphism and small-scale spatial variation in seedling density. *American Naturalist*, **146**, 685–707.

Godfray, H. C. J., Partridge, L. & Harvey, P. H. (1991). Clutch size. *Annual Review of Ecology and Systematics*, **22**, 409–429.

Haldane, J. B. S. & Jayakar, S. D. (1963). Polymorphism due to selection of varying direction. *Journal of Genetics*, **58**, 237–242.

Hartl, D. L. & Clark, A. G. (1989). *Principles of population genetics*. Sunderland, MA: Sinauer.

Holt, R. D. (1985). Population dynamics in two-patch environments; some anomalous consequences of an optimal habitat distribution. *Theoretical Population Biology* **28**, 181–208.

Holt, R. D. & Gaines, M. S. (1992). Analysis of adaptation in heterogeneous landscapes: implications for the evolution of fundamental niches. *Evolutionary Ecology* **6**, 433–447.

Houston, A. I. & McNamara, J. M. (1992). Phenotypic plasticity as a state dependent life-history decision. *Evolutionary Ecology* **6**, 243–253.

Howe, H. F. & Westley, L. C. (1986). Ecology of pollination and seed dispersal. In: *Plant ecology* (ed. M. J. Crawley). Oxford: Blackwell Scientific Publications.

de Jong, G. (1990). Quantitative genetics of reaction norms. *Journal of Evolutionary Biology* **3**, 447–468.

Kawecki, T. J. & Stearns, S. C. (1993). The evolution of life histories in spatially heterogeneous environments: optimal reaction norms revisited. *Evolutionary Ecology* **7**, 155–174.

Lechowicz, M. J. & Bell, G. (1991). The ecology and genetics of fitness in forest plants. II. Microspatial heterogeneity of the edaphic environment. *Journal of Ecology* **79**, 687–696.

Levene, H. (1953). Genetic equilibrium when more than one niche is available. *American Naturalist* **87**, 331–333.

Levins, R. (1962). Theory of fitness in a heterogeneous environment. I. The fitness set and adaptive function. *American Naturalist* **96**, 361–373.

Levins, R. (1963). Theory of fitness in a heterogeneous environment. II. Developmental flexibility and niche selection. *American Naturalist* **97**, 75–90.

Levins, R. (1968). *Evolution in changing environments. Some theoretical explorations.* Princeton University Press.

McGinley, M. A., Temme, D. H. & Geber, M. A. (1987). Parental investment in offspring in variable environments: theoretical and empirical considerations. *American Naturalist* **130**, 370–398.

Metz, J. A. J., Nisbet, R. M. & Geritz, S. A. H. (1992). How should we define 'fitness' for general ecological scenarios? *Trends in Ecology and Evolution* **7**, 198–202.

Mitchell-Olds, T. (1992). Does environmental variation maintain genetic variation? A question of scale. *Trends in Ecology and Evolution* **7**, 397–398.

Pulliam, H. R. (1988). Sources, sinks, and population regulation. *American Naturalist* **132**, 652–661.

Pulliam, H. R. & Danielson, B. J. (1991). Sources, sinks, and habitat selection: a landscape perspective on population dynamics. *American Naturalist* **137**, S50–S66.

Ricklefs, R. E. (1990). *Ecology.* New York: W. H. Freeman.

Roff, D. A. (1992). *The evolution of life histories: theory and analysis.* London: Chapman & Hall.

Schaffer, W. M. (1974a). Selection for optimal life histories: the effects of age structure. *Ecology* **55**, 291–303.

Schaffer, W. M. (1974*b*). Optimal reproductive effort in fluctuating environments. *American Naturalist* **108**, 783–790.

Sibly, R. M. (1995). Life-history evolution in spatially heterogeneous environments, with and without phenotypic plasticity. *Evolutionary Ecology* **9**, 242–257.

Sibly, R. M. & Antonovics, J. (1992). Life-history evolution. In: *Genes in ecology* (ed. R. J. Berry, T. J. Crawford & G. M. Hewitt), pp. 87–122. Oxford: Blackwell Scientific Publications.

Sibly, R. M. & Calow, P. (1986). *Physiological ecology of animals*. Oxford: Blackwell Scientific Publications.

Sibly, R. M. & Curnow, R. N. (1993). An allelocentric view of life-history evolution. *Journal of Theoretical Biology*, **160**, 533–546.

Sibly, R. M., Linton, L. & Calow, P. (1991). Testing life-cycle theory by computer simulation – II. Bet-hedging revisited. *Computer in Biology and Medicine* **21**, 345–355.

Stearns, S. C. (1976). Life-history tactics: a review of the ideas. *Quarterly Review of Biology* **51**, 3–47.

Stearns, S. C. (1992). *The evolution of life histories*. Oxford: Oxford University Press.

Stearns, S. C. & Koella, J. (1986). The evolution of phenotypic plasticity in life-history traits: predictions for norms of reaction for age- and size-at-maturity. *Evolution* **40**, 893–913.

Stratton, D. A. (1994). Genotype-by-environment interactions for fitness of *Erigeron annuus* show fine-scale selective heterogeneity. *Evolution* **48**, 1607–1618.

Tuljapurkar, S. D. (1982). Population dynamics in variable environments. II. Correlated environments, sensitivity analysis and dynamics. *Theoretical Population Biology* **21**, 114–140.

Tuljapurkar, S. D. (1990). *Population dynamics in variable environments*. New York: Springer-Verlag.

van Tienderen, P. H. (1991). Evolution of generalists and specialists in spatially heterogeneous environments. *Evolution* **45**, 1317–1331.

Via, S. & Lande, R. (1985). Genotype-environment interaction and the evolution of phenotypic plasticity. *Evolution* **39**, 505–522.

Wallace, B. (1968). *Topics in population genetics*. New York: W. W. Norton.

Appendix

Proof that, in each habitat, the optimal plastic strategy maximizes reproductive value at birth, calculated using global fitness.

Let $\underline{b}(h)$ represent the life history in habitat h of the carriers of a particular allele. Let $\underline{B}$ represent the set of life histories of the carriers of the allele, i.e. the set of $\underline{b}(h)$s for all habitats. The fitness of the allele, designated F, is defined by eqn (2), i.e.:

$$1 = \sum_{h,i} q(h) e^{-Ft_i} b_i(h) \tag{A1}$$

Note that to each $\underline{B}$ there corresponds one and only one value of F. The optimal $\underline{B}$, $\underline{B}^*$, is that maximizing F in eqn (A1). The habitat-specific life histories comprising $\underline{B}^*$ will be described as optimal life histories, and, individually, these will be designated $\underline{b}^*(h)$. The maximum value of F will be written F^*. In other words F^* by definition satisfies the inequality:

$$F^* \geqslant F \tag{A2}$$

for all possible $\underline{B}$s.

Reproductive value at birth in habitat h when the life history is $\underline{b}(h)$ and fitness is F is defined as:

$$V(h,\underline{b}(h),F) = \sum_{i} e^{-Ft_i} b_i(h) \tag{A3}$$

Note that $V(h, \underline{b}(h), F)$ is a monotonic decreasing function of F, and that from (A1),

$$\sum_{h} q(h) V(h,\underline{b}(h),F) = 1$$

We now prove the main theorem.

Theorem: The optimal life history in habitat h maximizes reproductive value at birth in that habitat, i.e. $\underline{b}^*(h)$ maximizes $V(h, b(h), F)$, where it is assumed that, outside habitat h, optimal life histories are used.

Proof: We seek to show there does not exist a life history $\underline{b}'$ in habitat h with the property that:

$$V(h,\underline{b}'(h),F') > V(h),\underline{b}^*(h),F^*) \tag{A4}$$

where F' is the fitness associated with $\underline{b}'(h)$ when in habitats other than h, the optimal life history is used.

From (A2), $F^* \geqslant F$, and since reproductive value at birth in habitat h is a monotonic decreasing function of F, it follows that:

$$\sum_{h' \neq h} q(h')V(h',\underline{b}^*(h'),F') \geqslant \sum_{h' \neq h} q(h')V(h',\underline{b}^*(h'),F^*) \tag{A5}$$

and from (A4):

$$q(h)V(h,\underline{b}'(h),F') > q(h)V(h,\underline{b}^*(h),F^*) \tag{A6}$$

Combining (A5) and (A6)

$$\sum_{h' \neq h} q(h')V(h',\underline{b}^*(h'),F') + q(h)V(h,\underline{b}'(h),F')$$

$$> \sum_{h'} q(h')V(h',\underline{b}^*(h'),F^*) \tag{A7}$$

By eqn (A1) the right-hand side of eqn (A7) equals 1, so the left-hand side is greater than 1. This, however, contradicts the definition of F', which is that the left-hand side of eqn (A7) equals 1 (from eqn (A1)). Hence the supposition that there exists a life history satisfying (A4) leads to a contradiction. It follows that no such life history exists, which is what we set out to prove.

V • Interactions

13 • Insect–plant interactions: the evolution of component communities

Douglas J. Futuyma and Charles Mitter

1 Component communities, plant life histories and insect diets

Root (1973) designated a plant species and the species associated with it, such as herbivores and their predators, a 'component community'. We will restrict our use of the term to the herbivorous arthropods. Such a component community varies with geography and ecological conditions, and overlaps with other plant species' component communities to the extent that the insects have broader or narrower host ranges.

In this chapter, we sketch some of our attempts to understand the influences on the diversity and identity of the species of herbivorous insects in plants' component communities. We emphasize the evolution of insect–host associations, but our questions bear on the evolution of plant life histories to the extent that the composition of a component community affects the evolution of a plant's defences, phenology and demographic characters. Herbivory has been postulated by many authors to have affected the evolution of flowering, fruiting and vegetative phenology, masting, age at first reproduction, and a host of morphological and chemical characters, and some (all too few) of these hypotheses have been supported by experimental evidence (Marquis 1992). Costs of resistance (preventing damage by insects) and of tolerance (compensating for tissue lost to herbivory) may affect allocation to growth and reproduction (Simms 1992). Several authors, for example, have postulated that allocation to defence is greatest in slowly growing plant species and those that grow in resource-poor environments (Janzen 1974, Coley *et al.* 1985).

The diversity and identity of a plant's herbivore fauna, rather than simply its biomass and the toll taken in reproductive output, will affect the evolution of defences to the degree that different defences are required to combat different enemies. A broadly effective defence will have correlated effects – perhaps measured by a positive genetic correlation in resistance – on different insects. Conversely, defences might have negatively correlated effects if a feature that deters one herbivore species is an attractant or feeding stimulant to another. Instances of both positive and negative correlations have been described, but most of the few reported genetic correlations in resistance to different herbivore species have not differed significantly from zero (Rausher 1992). This is puzzling, in view of the taxonomically broad deterrent and/or antibiotic effects of many secondary compounds in laboratory tests, but since genetic correlation studies generally use whole plants, the low genetic correlations in resistance may signal idiosyncratic, species-specific responses of different insects to multiple plant characters (Futuyma & May 1991). Moreover, traits that confer generalized resistance might arise rarely but become fixed rapidly, eroding the variation necessary for intraspecific study. In any case, present evidence suggests that the efficacy of defensive characters depends on the identity of the herbivore species, so the composition of a component community should affect the course of plant evolution.

What determines the diversity and composition of a plant's fauna? The many factors include plant architecture, breadth of the species' geographic distribution and the duration of residence in a region (Strong et al. 1984). At least some introduced plant species are attacked by fewer insect species than in their region of origin (Roeske et al. 1976; Zwölfer 1988). Thus history plays a role; but how deep a history must be taken into account? Specifically, what role does phylogenetic history play, compared with rapid, recent adaptation, in determining a plant's insect fauna? This is the major focus of our efforts.

Systematic entomologists have long known that in many taxa, related insects tend to feed on related plants – they are phylogenetically conservative in diet. Ehrlich & Raven (1964) summarized many such examples in the larval diets of butterflies, and used these as a foundation for their speculative scenario of long-term coevolution. They postulated that (a) insects' diets are determined largely by deterrent, attractant and toxic plant secondary compounds; (b) selection by insects leads to the evolution of new plant defences; (c) occasional evolution of a truly novel or efficacious chemical defence may liberate a plant lineage from most of its herbivores and foster diversification of the lineage in a relatively enemy-free 'adaptive zone'; (d) insect species associated with other plants eventually adapt to the radiating plant clade and themselves then diversify, occupying the diverse resources that these plants present.

Without evaluating the validity or likelihood of all points in this scenario, we aim in our work (1) to develop quantitative assessments of the importance of phylogeny in insect host associations, which is a conspicuous feature of

Ehrlich & Raven's model; and (2) to assess the possible role of genetic constraints as factors in the phylogenetic conservatism of diet. That is, has the history of host associations been guided not only by ecological sources of selection, but also by available genetic variation?

2 Phylogenetic conservatism of host-plant choice

Phylogenetic conservatism in host use might be assessed in several ways. In the absence of absolute datings, which is typical for insect/host associations, the rate or frequency of change in host choice is most usefully compared with that of cladogenesis, by mapping host use on insect phylogenies inferred from other characters. This approach has been used to ask how frequently the divergence of sister species has been accompanied by a difference in host family, often used as a criterion of 'major' host shift (Strong *et al.* 1984), as the great majority of individual insect herbivore species are restricted to plants of a single family (Bernays & Chapman 1994). An initial survey of 25 insect groups ranging from subgeneric to subfamily rank (Mitter & Farrell 1991), chosen solely because a cladogram was available, suggested that on average, sister species retain the same host family about 80% of the time (range 33–100%). In nine of these clades, which contain from eight to 126 species, there is no variation at all in host family use. Conservatism at this level probably represents a preponderance of small changes in the evolution of traits controlling diet, rather than complete stasis, because related insect species very often use different confamilial host species or genera. For example, although all 13 species of *Ophraella* leaf beetles are restricted to Asteraceae, nearest relatives seldom feed on the same plant genus (Funk *et al.* 1995).

In a few instances, absolute ages can be plausibly assigned to such conserved host associations. Usually these estimates are indirect. Thus, fossils are lacking for the lepidopteran suborder Heterobathmiina, comprising about ten extant congeners in southern South America, all feeding on *Nothofagus* (Fagaceae; Humphries *et al.* 1986). From its cladistic position with respect to known fossils of other lineages, however, *Heterobathmia* can be inferred to date from the Lower Cretaceous or earlier (125 million years ago; Labandeira *et al.* 1994), while *Nothofagus* is known from the Upper Cretaceous (Romero 1986). Thus, this lineage of moths has perhaps been restricted to the same host genus for at least 70 million years. Fossils of endophagous herbivory in which both the host and the herbivore can be identified, particularly leaf mines, are beginning to provide more direct evidence for long duration of host association. Labandeira *et al.* (1994) describe fossil leaf mines from the mid-Cretaceous (97 million years ago) which are attributable to the extant nepticulid moth genus *Ectodemia*. The hosts are undescribed relatives of modern Platanaceae, and the mine is described as 'remarkably similar' to that of an extant congener that feeds on *Platanus*.

Comparison of insect and host-plant phylogenies provides another kind of clue on evolutionary stability of host choice. In conjunction with other evidence (Mitter & Farrell 1991), significant match between such phylogenies can be taken to indicate that the insects have diversified in tandem with the host lineage, allowing the age of the insects' host preferences to be inferrred from that of the hosts. Thus, the apparently closely parallel phylogenies of *Phyllobrotica* and related leaf beetles and their host-plants in the order Lamiales allows this association to be reasonably dated to the mid-Tertiary from the fossil record of lamialean pollen, although fossils for the beetles are lacking (Farrell & Mitter 1990). Close concordance of phylogenies on this scale appears to be rare (Mitter & Farrell 1991), but approximate match, over much longer time spans, may be more widespread. For example, among the basal (non-ditrysian) lineages of Lepidoptera, the sole conifer feeders, Agathiphagidae (two species) are sister group to nearly all the rest, including all groups feeding ancestrally and predominantly on angiosperms (Nielsen 1989). Similarly cladistically basal, species-poor groups feeding on 'gymnosperms' are characteristic of other holometabolous herbivore groups of similar or greater age, such as Hymenoptera (sawflies: Naumann 1991) and, among the beetles, Chrysomeloidea (leaf beetles and allies: e.g. Kuschel & May 1990) and Curculionoidea (weevils; Anderson 1993). These apparently relictual groups have probably been associated continuously with 'gymnosperms' since before the Cretaceous rise of the angiosperms, which are now host to most herbivore species in these orders (review in Powell *et al.* 1997).

Some of the strongest evidence on conservatism and duration of host-taxon choice comes from comparison of relatives occurring in different biogeographic regions. (This approach was applied to plant demographic traits by Ricklefs & Latham 1992.) For example, Moran (1989) showed that the complex life cycle of melaphidine aphids alternates between mosses and sumacs (*Rhus*) in both eastern North America and eastern Asia, a geographic disjunction that may be more than 40 million years old. We have pursued this theme in a preliminary analysis of the diet of leaf beetles (Chrysomelidae), focussing on the 90 genera we know (from literature and examination of the Cornell University insect collection) to occur in New York State (NY).

Excluding genera with very polyphagous species or poorly documented diets (e.g. those in subfamily Donaciinae, which feed on aquatic plants, and Cryptocephalinae, which feed largely on decaying matter), we could identify 56 genera for which reasonably reliable host records could be found in the entomological literature for *both* at least one of the NY species and for *at least one* congeneric species in any of three other biogeographic regions: temperate Europe, western North America (wNA) (Rocky Mountains and west), and tropical America (Mexico and south). In the few cases in which NY beetles are classified as conspecific with congeners in other regions, we deleted host records of the species in those regions. Of the 56 genera, the NY representatives of 31 each feed on a single plant family, 14 are recorded from two families, and 11 from $\geqslant 3$ families. Europe harbours 27 of the NY genera, wNA 29 and tropical America 25. In aggregate, the hosts of these genera include 40 plant families in NY, 41 in Europe, 25 in wNA and 26 in tropical America, the latter two areas having been less studied.

There are several ways of characterizing the similarity of host associations across different regions. For instance, we asked, of the 27 NY genera shared with Europe, what fraction use at least one plant family that is also the host of European congeners. The answer is apparently 26/27: 15 of 16 genera recorded from one plant family in NY use the same family in Europe, and all of the 11 that use two or more families in NY feed on at least one of these families in Europe. We calculated the expected number of such matches from the probability of drawing at random the family (families) used by each genus in NY from the 41 plant families used in Europe by congeners of all NY beetles. The expected numbers of shared host associations, by this calculation, were far lower than the observed numbers (Table 13.1*a*). The incidence of shared use of plant families between NY and either wNA or tropical America was equally high and equally non-random (Table 13.1*a*). Table 13.1*b* includes the results of another approach to the data: calculating the mean proportion of host families used by a NY genus that are also hosts of congeners in each of the other regions. Especially for genera that in NY are restricted to one or two host families, these proportions are very high. We infer from this analysis that the chrysomelid fauna of NY – and of eastern North America, where many of the NY genera are broadly distributed – has beeen assembled, in part since the Pleistocene and undoubtedly in part before then, from lineages that invaded from other regions and which, almost to a beetle, retained their ancestral host-family associations. (The same conclusion applies to colonization in the reverse direction.) That is not to deny that the eastern North American species have adapted to different species or genera of

Table 13.1. *Incidence of shared host-plant families among genera of Chrysomelidae in New York State and three other biogeographic regions*

(Chrysomelid genera are classified by number of plant families (k) used as hosts by New York species. In (a), the expected number of NY genera that share at least one plant family with congeners in another region (e.g. Europe) was calculated as GP_k, where G is the number of beetle genera in NY, k is the number of plant families in the diet of a NY genus, P_k is the probability that congeners in Europe include in their diet at least one of k specified plant families, K is the number of families in the diet of a genus in Europe, p_K is the proportion of genera in Europe that feed on K families, N is the number of plant families eaten by European congeners of NY beetles, f is the fraction of NY host families that occur in the diet of all European congeners, and $P_k = k\Sigma_K p_K f / N$.)

(a) Observed and expected numbers of New York genera sharing at least one host family with congeners in other regions

No. host families in NY (k):	1	2	$\geqslant 3$
Europe			
No. sharing/no. shared genera	15/16	9/9	2/2
Expected no. sharing	0.88	1.00	0.94
Western North America			
No. sharing/no. shared genera	11/15	6/6	6/7
Expected no. sharing	0.47	0.38	0.82
Tropical America			
No. sharing/no. shared genera	13/14	6/6	3/5
Expected no. sharing	0.46	0.39	1.11

(b) Mean proportion of host families of New York chrysomelid genera shared with congeners in other regions. The number of shared chrysomelid genera is in parentheses

No. host families used in NY:	1	2	3	$\geqslant 3$
Proportion shared in				
Europe	0.94 (16)	0.55 (9)	—	0.63 (2)
Western North America	0.81 (15)	0.50 (6)	0.33 (2)	0.34 (5)
Tropical America	0.93 (13)	0.75 (6)	0.00 (1)	0.22 (4)

plants – many have done so – but their adaptation has been taxonomically strongly circumscribed. A preliminary review of the chrysomelid fauna of the midwestern United States (specifically, that of Missouri; Riley & Enns 1979) pointed to a similar conclusion (Farrell & Mitter 1993).

3 Genetic variation in relation to host shifts

The phylogenetic conservatism of diet in many phytophage clades suggests the hypothesis that the evolution of host associations is guided not only by ecological sources of selection (e.g. rarity of the normal host, competition, host-associated predation), but also by 'internal' constraints that could be manifested as genetic correlations or limitations on genetic variation. That is, some conceivable host shifts may be more likely than others, for genetic reasons. Futuyma *et al.* (1993, 1994, 1995) have explored this hypothesis by screening for heritable variation in responses of several host-specialized species of *Ophraella* leaf beetles (Chrysomelidae) to host plants of their congeners. *Ophraella* is a North American genus, in which each of the 13 known species feeds on one or more species within one of four tribes of Asteraceae.

The genetic work has been pursued in the context of a provisional history of host shifts in the genus, inferred from the most parsimonious distribution of host taxa on a phylogeny of *Ophraella* based on mitochondrial DNA sequences (Figure 13.1). (This phylogeny is mostly congruent with an earlier estimate based on morphology and allozymes.) The *Ophraella* phylogeny is not highly congruent with that of the hosts, providing one of several lines of evidence for host-switching rather than cospeciation, but closely related species and populations frequently feed on plants in the same tribe, and in a few instances, cospeciation cannot be ruled out. The question was whether genetic variation might more frequently be detected in responses to either (a) plants closely related to a species' normal host, or (b) host plants of sister species, or otherwise closely related species, of beetles, representing host shifts realized during phylogeny. That is, might the presence vs. apparent lack of genetic variation predict the phylogenetic history of adaptation to new hosts?

In brief, evidence for heritable variation was sought as variance among half-sib or (in some cases) full-sib families in feeding responses of neonate larvae and newly eclosed adults to a plant species, and in larval survival. Animals were confined singly with discs of leaf tissue of one plant species, without choice, and the area consumed after 24 hours was measured. Leaf tissue was replaced periodically in survival tests. Variation in each of four species of *Ophraella* was scored with respect to four to six plant species that are hosts of *Ophraella* species other than the one tested.

Although all species provided evidence of genetic variation with respect to at least one test plant, we were unable to detect genetic variation in 14 of 16 tests of larval survival and in 18 of 39 tests of larval or adult consumption. In

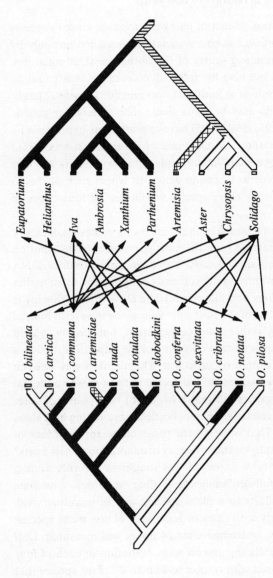

Figure 13.1. The estimated phylogenies of the leaf beetle genus *Ophraella* (left) and its host plants (right). Lines between the tips of the trees indicate host associations, and the shading of the *Ophraella* tree represents the most parsimonious history of evolutionary shifts between host tribes (open areas denote Astereae; cross-hatched areas denote Anthemideae; solid areas denote Eupatorieae; medium shaded areas denote Heliantheae), correspondingly shaded on the host-plant phylogeny (striped areas denote common ancestors). In a complete phylogeny of *Ophraella*, many tribes and genera would be inserted among those shown. The *Ophraella* phylogeny is based on mitochondrial DNA sequences, and that of the plants chiefly on RFLP and sequence analysis of chloroplast DNA; see Funk *et al.* (1995) for methods and references.

Table 13.2. *Tests for association between detected (+) and undetectable (−)
genetic variation in consumption, by larval or adult* Ophraella, *of hosts of their
congeners, and each of several aspects of plant or* Ophraella *phylogeny*

(One-tailed exact probabilities obtained by likelihood-ratio tests (STATEXACT
package)[a].)

	−	N	p
(a) Test plant in same or different tribe as host of beetle species scored				
Same	7	1	0.0373
Different	14	17	
(b) Test plant represents or not an immediate host shift on *Ophraella* phylogeny, with respect to beetle species scored				
Does	6	3	0.3116
Does not	15	15	
(c) Test plant is, or is not, host of an *Ophraella* species in the same major clade as the species scored.[b]				
Is	12	4	0.0288
Is not	9	14	

[a] Modified from Futuyma *et al.* (1995).
[b] The major clades (see Figure 13.1) are (1) *pilosa*; (2) *conferta–notata*; (3) *bilineata–slobodkini*.

some but not all of the negative cases, virtually all individuals refused to feed
and soon died. These results are strikingly at variance with most studies of
Drosophila and other organisms, in which most characters display heritable
variation. Although clearly we can never say that genetic variation in a
character is absolutely lacking, the evident differences in magnitude of genetic
variation in responses to different plants suggest greater potential for adaptation to some than to others.

In both of the two cases in which genetic variation in larval survival was
detected, the test plant was in the same subtribe (Ambrosiinae) as the beetle's
normal host. We used likelihood ratio tests for association between the
presence vs. 'absence' of genetic variation in consumption, and several aspects
of plant and insect phylogeny. Plants in the same tribe of Asteraceae as a
species' normal host were significantly more likely to evoke genetically variable consumption than plants in different tribes (Table 13.2a). No association
was found between genetic variation and whether or not the test plant
represented an immediate host shift in the *Ophraella* phylogeny. But genetic
variation was more frequently manifested in consumption of hosts of closely

related *Ophraella* species (those in the same major clade) than hosts of more distantly related species (Table 13.2c). Because closely related *Ophraella* species generally have hosts in the same tribe of Asteraceae, we cannot tell whether genetic variation is more strongly correlated with plant relationships or beetle relationships. Perhaps, though, the distinction is unimportant; what is important is that the genetic data, *mirabile dictu*, are consistent with the major feature of phylogenetic history, namely that shifts among closely related plants are more frequent than among distantly related plants.

4 Conclusion

Taxonomic, phylogenetic, and now genetic evidence are consistent in affirming that in many (but not all) groups of host-specialized insects, host shifts occur most frequently among closely related plants, perhaps due to biases in available genetic variation. Whether we should be surprised by these constraints and by the conservatism of insect diet, or should simply take them as one more instance of Darwinian gradualism, of *Natura non facit saltum*, perhaps depends on what constraints we imagine might inhere in such features as chemoreception, central processing of sensory input and detoxification of plant compounds. Since very little is known about these mechanisms, much less their capacity for variation, we can say little about what constraints to expect.

We do not wish to imply that phylogenetic history is all-important in explaining insect/plant associations. It does not account for the associations of highly polyphagous species or of the fairly numerous instances of closely related species with distantly related hosts. But it is nonetheless clear that evolutionary history, extending back many millions of years, leaves a significant imprint on the composition of the component communities that must exert selection on plant characteristics. These patterns speak of opportunitiy for prolonged, consistent selection imposed by long-term associations, perhaps resulting in coevolved adaptation; and, conversely, of possible shifts in or relaxation of selection on plants that have escaped their specialized herbivores by colonizing different regions. Evolutionary ecologists have only begun to address such questions.

References

Anderson, R. S. (1993). Weevils and plants: phylogenetic versus ecological mediation of evolution of host associations in Curculioninae (Coleoptera: Curculionidae). *Memoirs of the Entomological Society of Canada* **165**, 197–232.

Bernays, E. A. & Chapman, R. F. (1994). *Host-plant selection by phytophagous insects*. New York: Chapman & Hall.

Coley, P. D., Bryant, J. P. & Chapin, F. S. III. (1985). Resource availability and plant antiherbivore defense. *Science* **230**, 895–899.

Ehrlich, P. R. & Raven, P. H. (1964). Butterflies and plants: a study in coevolution. *Evolution* **18**, 586–608.

Farrell, B. & Mitter, C. (1990). Phylogenesis of insect/plant interactions: have *Phyllobrotica* and the Lamiales diversified in parallel? *Evolution* **44**, 1389–1403.

Farrell, B. & Mitter, C. (1993). Phylogenetic determinants of insect/plant community diversity. In *Species diversity in ecological communities: Historical and geographic perspectives* (ed. R. E. Ricklefs & D. Schluter), pp. 253–267. Chicago: University of Chicago Press.

Funk, D. J., Futuyma, D. J., Ortí, G. & Meyer, A. (1995). A history of host associations and evolutionary diversification for *Ophraella* (Coleoptera: Chrysomelidae): new evidence from mitochondrial DNA. *Evolution* **49**, 1008–1017.

Futuyma, D. J. & May, R. M. (1992). The coevolution of plant/insect and parasite/host relationships. In *Genes in ecology* (ed. R. J. Berry, T. J. Crawford & G. M. Hewitt), pp. 139–166. London: Blackwell.

Futuyma, D. J., Keese, M. C. & Scheffer, S. J. (1993). Genetic constraints and the phylogeny of insect-plant associations: responses of *Ophraella communa* (Coleoptera: Chrysomelidae) to host plants of its congeners. *Evolution* **47**, 888–905.

Futuyma, D. J., Walsh, J., Morton, T., Funk, D. J. & Keese, M. C. (1994). Genetic variation in a phylogenetic context: responses of two specialized leaf beetles (Coleoptera: Chrysomelidae) to host plants of their congeners. *Journal of Evolutionary Biology* **7**, 127–146.

Futuyma, D. J., Keese, M. C. & Funk., D. J. (1995). Genetic constraints on macroevolution: the evolution of host affiliation in the leaf beetle genus *Ophraella*. *Evolution* **49**, 797–809.

Humphries, C. J., Cox, J. M. & Nielsen, E. S. (1986). *Nothofagus* and its parasites: a cladistic approach to coevolution. In *Coevolution and systematics* (ed. A. R. Stone & D. L. Hawksworth), pp. 55–76. Oxford: Clarendon Press.

Janzen, D. H. (1974). Tropical blackwater rivers, animals, and mast fruiting by the Dipterocarpaceae. *Biotropica* **6**, 69–103.

Kuschel, G. & May, M. M. (1990). Palophaginae, a new subfamily for leaf-eating beetles, feeding as adult and larva on araucarian pollen in Australia (Coleoptera: Megalopodidae). *Invertebrate Taxonomy* **3**, 697–719.

Labandeira, C. C., Dilcher, D. L., Davis, D. R. & Wagner, D. L. (1994). Ninety-seven million years of angiosperm-insect association: Paleobiological insights into the meaning of coevolution. *Proceedings of the National Academy of Sciences of the U.S.A.* **91**, 12278–12282.

Marquis, R. J. (1992). Selective impact of herbivores. In *Plant resistance to herbivores and pathogens: ecology, evolution, and genetics* (ed. R. S. Fritz & E. L. Simms), pp. 301–325. Chicago & London: The University of Chicago Press.

Mitter, C. & Farrell, B. (1991). Macroevolutionary aspects of insect–plant relationships. In *Insect–plant interactions*, vol. III (ed. E. A. Bernays), pp. 36–78. Boca Raton, FL: CRC Press.

Moran, N. A. (1989). A 48-million-year-old aphid-host plant association and complex life cycle: biogeographic evidence. *Science* **245**, 173–175.

Naumann, I. D. (1991). Hymenoptera. In *The insects of Australia*, 2nd edn, vol. 2 (ed. Commonwealth Scientific and Industrial Research Organisations), pp. 916–1000. Carlton: Melbourne University Press.

Nielsen, E. S. (1989). Phylogeny of major lepidopteran groups. In *The hierarchy of life* (ed. B. Fernholm *et al.*), pp. 281–294. Amsterdam: Elsevier Science Publ.

Powell, J. A., Mitter, C. & Farrell, B. D. (1996). Evolution of larval feeding habits in Lepidoptera. In *Handbook of zoology. Lepidoptera*, vol. 1, *Systematics and evolution* (ed. N. P. Kristensen) Berlin: Walter de Gruyter. (In the press.)

Rausher, M. D. (1992). Natural selection and the evolution of plant–insect interactions. In *Insect chemical ecology: an evolutionary approach* (ed. B. D. Roitberg & M. B. Isman), pp. 20–88. New York & London: Chapman & Hall.

Ricklefs, R. E. & Latham, R. E. (1992). Intercontinental correlation of geographical ranges suggests stasis in ecological traits of relict genera of temperate perennial herbs. *American Naturalist* **139**, 1305–1321.

Riley, E. G. & Enns, W. R. (1979). An annotated checklist of Missouri leaf beetles (Coleoptera: Chrysomelidae). *Transactions of the Missouri Academy of Sciences* **13**, 53–82.

Roeske, C. M., Seiber, J. N., Brower, L. P. & Moffitt, C. M. (1974). Milkweed cardenolides and their comparative processing by monarch butterflies. *Recent Advances in Phytochemistry* **10**, 93–159.

Romero, E. J. (1986). Fossil evidence regarding the evolution of *Nothofagus* Blume. *Annals of the Missouri Botanical Garden* **73**, 276–283.

Root, R. B. (1973). Organization of a plant-arthropod association in simple and diverse habitats: the fauna of collards (*Brassica oleracea*). *Ecological Monographs* **43**, 95–124.

Simms, E. L. (1992). Costs of plant resistance to herbivory. In *Plant resistance to herbivores and pathogens: ecology, evolution, and genetics* (ed. R. S. Fritz & E. L. Simms), pp. 392–425. Chicago & London: The University of Chicago Press.

Strong, D. R., Jr., Lawton, J. H. & Southwood, T. R. E. (1984). *Insects on plants: community patterns and mechanisms*. Cambridge, MA: Harvard University Press.

Zwölfer, H. (1988). Evolutionary and ecological relationships of the insect fauna of thistles. *Annual Review of Entomology* **33**, 103–122.

14 · Evolutionary trends in root–microbe symbioses

A.H. Fitter and B. Moyersoen

1 Introduction

In common with most other organisms, the normal state of seed plants is symbiotic. They coexist with a range of symbionts, including leaf-inhabiting fungal endophytes, but the most widespread symbionts are in roots. Possibly 90% of all seed plant species have fungal symbionts in their roots, forming the structures known as mycorrhizas. A very much smaller number form nodules containing symbiotic bacteria that can fix atmospheric dinitrogen gas. There is strong evidence (discussed below) that the original colonization of land by plants was at least assisted by the evolution of mycorrhizal symbioses, but it is also certain that other types of symbiosis, both mycorrhizal and bacterial, have evolved subsequently. In this chapter, we aim to investigate the evolutionary history of these symbioses as it relates to the phylogeny of the seed plants, seeking any correlations between the development of different types of symbiosis and important ecological traits of the hosts. We shall concentrate on the mycorrhizal symbioses, because they are more varied and widespread than the other types.

Although the term mycorrhiza lacks a clear definition, it is best viewed as a sustainable, non-pathogenic, biotrophic interaction between a fungus and a root; this chapter will not cover pathogenic associations or the range of poorly-characterized associations that have been identified by mycologists, such as those involving 'dark septate hyphae' (Haselwandter & Read 1981; Stoyke et al. 1992) or dematiaceous fungi resembling *Phialocephala fortinii* (O'Dell et al. 1993). Mycorrhizas range widely in form and in the type of fungus involved, demonstrating that they do not represent a single evolutionary class of association, but rather a type of association that has evolved repeatedly, in response to distinct selection pressures. They have been classified in various ways: for example into ectomycorrhizas (EcM), typically formed between woody species and long-lived fungi (generally

Basidiomycotina and Ascomycotina), and endomycorrhizas. The latter themselves contain a number of disparate types, notably the ubiquitous arbuscular mycorrhizas (AM, sometimes referred to as vesicular-arbuscular mycorrhizas), involving a very wide range of plants and a small group of fungi in the Glomales (Zygomycotina); the ericoid mycorrhizas, formed between fungi in the Ascomycotina (notably *Hymenoscyphus* spp.) and plants in the Ericales; and the orchid mycorrhizas, formed between various fungi, often imperfect forms, and orchids. There are also numerous other types that have been wholly or partially described and are apparently confined to restricted plant taxa; they illustrate our imperfect understanding of this symbiosis. Functionally, these symbioses behave in distinct ways, but they share the transfer of fixed carbon from the plant to the fungus as a unifying physiological feature. The characteristics of the main types of mycorrhiza are summarized in Table 14.1.

In contrast, bacterial root symbioses are more sharply defined. There are two main types: (a) the rhizobial symbiosis is formed between most (but importantly not all) species of the Fabales and a single genus in the Ulmaceae (*Parasponia*), and bacteria in what was formerly the genus *Rhizobium*, now split into three or even four genera, including *Bradyrhizobium*, *Azorhizobium*, and recently *Sinorhizobium* (Young & Haukka, 1996); (b) actinorhizas develop between bacteria of the genus *Frankia* and an apparently diverse group of plants in 24 genera in eight families. In each case the bacteria develop within a nodule formed on the root and fix nitrogen within it, receiving fixed carbon from the plant, as in mycorrhizas.

2 The systematic pattern

The distribution of root microbe symbioses can be viewed under two heads: abundance and phylogenetic concentration. By far the most abundant of these symbioses is the arbuscular mycorrhiza. Trappe (1987) analysed available records of mycorrhizal colonization and found that 55% of angiosperm species examined (which was only 3% of all known angiosperm species) formed AM; a further 12% were regarded as facultatively mycorrhizal. Since this figure is probably biased towards preferential inclusion of non-mycorrhizal species such as annual weeds and crops, it can be estimated that around two-thirds of land plants normally form AM symbioses. Such figures have to be interpreted with great care. A sure demonstration that a plant species is 'normally mycorrhizal' ideally requires the following:

1. identification of characteristic structures (e.g. arbuscules in AM, Hartig net

Table 14.1. *The characteristics of the main types of mycorrhiza*

	Ectomycorrhiza	Arbuscular mycorrhiza	Ericoid mycorrhiza	Orchid mycorrhiza
Taxa				
Fungi involved	Basiodiomycotina, Ascomycotina, one genus in Zygomycotina	Zygomycotina (Glomales)	Ascomycotina (esp. *Hymenoscyphus*)	Basidiomycotina, Mycelia Sterilia, Deuteromycotina
Plants involved	taxonomically diverse, many	taxonomically diverse, few	Ericales	Orchidales
Morphology				
Diagnostic features	Hartig net	arbuscule	intra-cellular hyphal complexes in epidermis	hyphal coils (pelotons) in cortical cells
Other structures	sheath	vesicles, coils		
Functionality (changes in plant function)				
P acquisition	important	important	?important	exclusive[a]
N acquisition	important	marginal	important	exclusive
Water acquisition	?important	marginal	unknown	unknown
Protection from pathogens	?important	?important	unknown	unknown
Others	?protection from ionic toxins	?micronutrients	protection from ionic toxins	carbon supply to plant

[a] 'exclusive' is used to indicate that all the plant's resource acquisition is achieved by this route.

in EcM symbioses) in field-collected roots, and from a range of habitats if the species typically occurs in several;
2. re-synthesis of the association under controlled conditions, with appropriate plant and fungal taxa (i.e. Koch's postulates should be applied);
3. a demonstration of functional attributes of the association.

Unsurprisingly, this has been performed only for a handful of cases, and the vast bulk of the literature on mycorrhizal associations is based upon some subset of these criteria, with the first being the most frequently achieved. In many cases, however, even that most basic test is only partially satisfied, and records of arbuscular mycorrhizal associations without evidence of arbuscules, or of ectomycorrhizal associations without visible Hartig nets (both are diagnostic characters) abound. It is common, for example, to find apparently AM-colonized roots in which only hyphae and vesicles or coils are visible; experienced workers feel confident about identifying these as AM structures, even though they are not strictly diagnostic. More seriously, many of the older records of EcM associations are based on the presence around trees of carpophores of fungi known to be ectomycorrhizal, with no examination of roots at all. Even the basic data are therefore suspect; extrapolations from them must be treated with extreme caution.

Nevertheless, it is certain that the AM symbiosis is the most abundant type. It also shows the smallest degree of phylogenetic concentration, being found in almost all groups including the Bryopsida. Some of the symbioses demonstrate extreme concentration, occurring only in a single plant taxon, notably the orchid and ericoid mycorrhizas, and (with the sole exception of *Parasponia*) the rhizobial nodule symbiosis. To some extent, this is a circular argument: there may be a tendency for investigators to describe a symbiosis found in such a group as being of the type defined by that group, but even so the statement appears robust. The other two types to be discussed here, actinorhizas and ectomycorrhizas, are often cited as cases where there is no apparent phylogenetic pattern to their distribution. Actinorhizas are found in eight families including Betulaceae, Casuarinaceae, Myricaceae and Rosaceae, while ectomycorrhizas appear in families as disparate as the Pinaceae, Fabaceae, Dipterocarpaceae and Cyperaceae.

2.1 The ancestral symbiosis

Pirozynski & Malloch (1975) were the first to suggest not only that the evolution of the arbuscular mycorrhizal symbiosis was contemporaneous with the first land plants, but that symbiosis between a photosynthetic aquatic ancestor and a fungus was actually a prerequisite for the development of a land flora. The argument was largely based on the inability of essentially rootless plants to acquire sufficient phosphorus: the phosphate ion is virtually immobile in soil (phosphate ions will diffuse 0.1 to 1 mm per day in typical soils) and its acquisition requires the external hyphae of the fungal partner in the symbiosis. Other important evidence was the remarkable ubiquity of the symbiosis in the plant kingdom, and the tentative identification of structures in 400 million year old Early Devonian fossils (*Aglaophyton*) from the Rhynie Chert as being hyphae and vesicles from an AM fungus (Kidston & Lang, 1921).

This hypothesis has recently received convincing confirmation from two additional pieces of evidence. First, molecular sequence data have been used (Simon *et al.* 1993) to estimate the origin of the order Glomales, which includes all known AM fungi, at between 353 and 462 million years ago, consistent with the believed origin of a land flora. Second, Taylor *et al.* (1995) have re-examined the *Aglaophyton* material and shown that the endophytic fungus produced arbuscules, confirming its AM nature. It seems certain, therefore, that the AM symbiosis developed shortly after or even simultaneously with the appearance of a land flora, and Pirozynski & Malloch's (1975) perceptive theory appears well-founded. This means that ancestors of all land plants had the potential to form arbuscular mycorrhizas, and those that do not do so nowadays have either lost or suppressed the genes involved. Evidence that the genes are not lost in many cases comes from the widespread occurrence of dual symbioses (typically AM and EcM in the same root) and from the recent discovery that many species in the Pinaceae are naturally colonized by AM fungi in the wild (Cázares & Trappe 1993; Cázares & Smith 1996).

From the standpoint of comparative analysis, the questions that need to be asked, therefore, are not about the distribution of the AM symbiosis and its ecological correlates, but about the pattern of the other symbioses and, importantly, of the 10% or so of plant species which appear to have abandoned this symbiosis altogether.

3 Phylogenetically concentrated symbioses

3.1 Orchid and ericoid mycorrhizas

These two symbioses both occur within a single family (Orchidaceae) or order (Ericales), as far as is known, and it is reasonable to postulate that the evolutionary event that led to their development occurred early in the establishment of the respective taxa. However, not all species within these taxa necessarily form the eponymous symbiosis: there are records of AM associations in orchids (Hall 1976) and certainly in other families of the Orchidales, such as Burmanniaceae (Leake 1994). The Ericales includes families such as the Pyrolaceae and Monotropaceae which have their own distinctive mycorrhizal types. Nor is it the case that the fungus partner in these symbioses shows a high degree of faithfulness: the orchid fungi are typically abundant species that may even be pathogens of trees, and Duckett & Read (1995) have shown that the ericoid endophyte *Hymenoscyphus ericae* can colonize the rhizoids of a number of leafy liverworts; whether the liverwort association is truly mycorrhizal awaits confirmation.

The orchid association appears to be unique in that all identified material fluxes in the symbiosis are from fungus to host: in other words the orchid appears to parasitize the fungus. This is transparently true of the achlorophyllous orchid genera such as *Neottia* and *Epipogium* which have neither chlorophyll nor roots and are therefore incapable of autonomous resource acquisition. But even in green orchids such as *Goodyera repens*, no carbon movement from plant to fungus was seen by Alexander & Hadley (1984). This extreme mycotrophy, associated with achlorophylly, has also evolved in two families of the Ericales, the Monotropaceae and the Pyrolaceae (reduced by some taxonomists to subfamilies of the Ericaceae). A phylogeny of the Ericales based on 28S rRNA sequences (Cullings 1994) shows that the mycorrhizal types are congruent with the host phylogeny, to the extent that the subfamily Monotropoideae (syn. Family Monotropaceae) is both polyphyletic on morphological grounds, and contains more than one mycorrhizal type (Figure 14.1). *Monotropsis ordata* appears in the phylogeny in subfamily Vaccinioideae (third from top in Figure 14.1) and is the only member originally classified as Monotropoideae not to have a typical monotropoid mycorrhiza. Cullings *et al.* (1996) have shown that some monotropoid mycotrophs, including *Monotropa hypopithys* and, especially, *Pterospora andromeda*, are specialist parasites of a small number of EcM associations, and that such specialization has developed within in a single clade of the Monotropoideae.

The complexity of mycorrhizal types in the Ericales is greater than in any other order: at least four distinct types have been recognized, including ectomycorrhiza, ericoid, arbutoid and monotropoid mycorrhiza (see Harley & Smith 1983 for details of these associations). Each of these is defined morphologically but named for the particular plant taxon involved. Molecu-

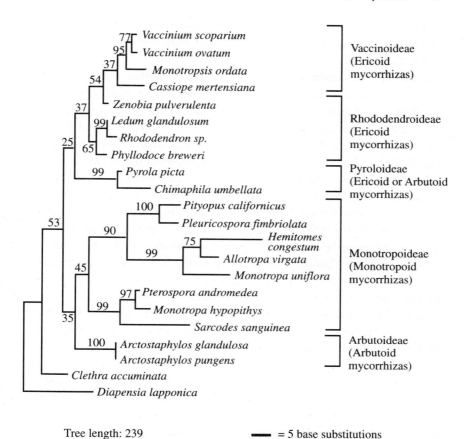

Tree length: 239   ▬▬ = 5 base substitutions
C.I.: 0.680

Figure 14.1. Phylogenetic tree of the Ericales based upon 28S rRNA gene sequences, showing the conformity of phylogeny and mycorrhizal type (Cullings 1994).

lar analyses confirm that the mycorrhizal status of a plant species is an important taxonomic character within this group: in other words, the nature of the mycorrhiza is a determinant of ecological and evolutionary behaviour.

3.2 Nitrogen-fixing nodules

The rhizobial symbiosis has a curious distribution, with most members of two subfamilies of the Fabaceae (Papilionoideae and Mimosoideae) normally forming the symbiosis, but only about a third of the species in the third subfamily, Caesalpinioideae. In addition, the single genus *Parasponia* (Ulmaceae) has also been shown to be rhizobial. This curious pattern has provoked speculations of unusual evolutionary phenomena such as horizontal transfer.

In contrast, the actinorhizal symbiosis has been viewed as phylogenetically diffuse. Associations have been found in 24 genera in eight families. In traditional angiosperm classifications, notably that of Cronquist (1981), this distribution showed no obvious pattern. Seven orders were involved: Rhamnales, Proteales and Rosales in the Rosanae; Casuarinales, Myricales and Fagales in the Hammamalidanae; and Ranunculales in the Magnolianae. A quite different picture emerges using a different phylogeny, based on DNA sequences of the chloroplast *rbc*L gene (Chase *et al.* 1993). This suggests that all N-fixing root symbioses, both rhizobial and actinorhizal, occur in a single clade, rosid I (Soltis *et al.* 1995), implying a single origin of the potential to form the symbioses, followed either by the development of distinct types of symbiosis in some subsequent lineages, or possibly the loss of the symbiosis in others. Whichever is true, it opens the possibility of asking comparative questions about the species in the rosid I clade.

The traditional adaptationist view of the N-fixing symbioses has been that they permit species to grow in N-deficient habitats. This has always been hard to maintain. Some N-fixers do indeed occur in, for example, early successional habitats: the well-documented case of *Alnus crispa*, an actinorhizal species, in the succession at Glacier Bay, Alaska, is a good illustration. But many legumes, in particular, are not obviously associated with extreme N-deficiency. An alternative view, put forward by McKey (1994), is that the legumes (Fabaceae) were a group with inherently high nitrogen demand, characterized by high leaf tissue N concentrations and consequent high photosynthetic rates (Field & Mooney 1986). As a result, a symbiosis that increased N acquisition ability would be strongly favoured.

If this view is accepted, the interesting question becomes why some legumes are non-rhizobial. In Caesalpinoideae (Fabaceae), the nodulating tribes Caesalpiniae and Detariae are predominantly AM, whereas ectomycorrhizal associations are only frequent in the tribe Amherstiae, members of which do not normally nodulate (Alexander 1989). Possibly the carbon cost of maintaining both symbioses is too great, though certainly some species (e.g. the actinorhizal *Alnus* and the rhizobial *Acacia*) can achieve such multiple symbiosis. In contrast, most Papilionoideae are both nodulated and arbuscular mycorrhizal: the carbon cost of AM symbiosis is almost certainly less than that of EcM. An alternative explanation for the paucity of EcM nodulators is that the same genes are required, as is true of *nod⁻* peas, in which both nodulation and AM formation are suppressed (Ginaninazzi-Pearson *et al.* 1996).

A wider view of the evolution of nodulation symbioses might allow a range of selection pressures to have acted within a clade that possessed the requisite genetic mechanisms. Doyle (1994) suggests that not only is the sub-family Caesalpinoideae polyphyletic, but that nodulation has arisen several times within the family. In some cases the driving force may have been a suite of characters that demanded high rates of N acquisition; in others, both legumes and actinorhizal non-legumes, nodulation may have been favoured in taxa that occupied N-deficient habitats.

Actinorhizal symbioses share some striking phylogenetic patterns with some mycorrhizas, notably a match between the phylogeny of the plant and microbial partners. This matching has been described in ericoid mycorrhizas (see above) and also in actinorhizas (Cournoyer *et al.* 1993). It seems likely that these symbioses have a powerful effect on the ecology and consequently constrain the evolution of both partners.

4 Phylogenetically diffuse symbioses

Where symbioses have evolved within well-defined clades and are character-istic of most members of that clade (e.g. orchid mycorrhizas), there is little scope for comparative analysis. In some cases, such as N-fixing symbioses, the patterns are much more amenable to analysis, but current uncertainties about phylogenetic relationships or lack of basic biological and ecological information about the species makes progress difficult. There are, however, two groups of associations that offer more scope, even with current data, namely ectomycorrhizal and (stretching a point) non-mycorrhizal species.

4.1 Ectomycorrhizas

On the basis of the Cronquist (1981) classification, EcM species of plant are found in all the major groups of dicots and in one group of monocots, the Cyperales. Trappe (1987) has already presented a comprehensive analysis of the phylogenetic trends that can be detected in this way. He suggested that the most pronounced evolutionary trend was towards a reduced dependence on mycorrhizal associations, and also noted the absence of mycorrhizal types other than AM in annual plants. There is a parallel association with woodi-ness: virtually all EcM plant species are woody (trees, shrubs, lianes), as shown by the results of a principal components analysis of angiosperm

Table 14.2. *Loadings on the first two axes of a*
principal components analysis of orders of angiosperms
(Each variable refers to the presence of at least one
record of that characteristic in the order. The first two
axes accounted for 49% of the variation.)

Variable	Axis 1	Axis 2
Eigenvalue	2.091	1.319
Ectomycorrhiza	0.593	−0.089
Dual mycorrhiza (AM/EcM)	0.611	−0.007
Non-mycorrhizality	−0.064	0.584
Other mycorrhizal types	−0.178	−0.310
Nitrogen fixation	0.436	−0.188
Parasitism (plant–plant)	0.025	0.528
Insectivory	0.220	0.491

orders, based upon the presence and absence of various types of symbiosis (Table 14.2). The first component in this PCA has large loadings for EcM, dual (EcM plus AM) mycorrhiza and N-fixation. An analysis of variance of the first axis scores of each order using the presence or absence of woody species in the order as the factor, reveals that non-woody orders have negative scores (− 0.867 ± 0.177; mean and 95% confidence limit) whereas orders with at least some woody species have positive scores (0.458 ± 0.432). The difference is highly significant: $F_{1,79} = 18.81$, $P < 0.001$. The association between woodiness and the occurrence of both EcM and N-fixation is therefore strong at the level of order.

There are a number of records of EcM in non-woody species, and these illustrate the severe difficulties of relying on published records for this type of analysis. Warcup (1980) reported that a number of Australian Asteraceae, including some annual herbs, could form ectomycorrhizas. Subsequently, Warcup & McGee (1983) tested numerous species of Asteraceae in culture with one or two ectomycorrhizal fungi and found that 20 annual species could apparently form EcM structures. The illustration they show of *Podolepis rugata* displays typical external morphology and also the Hartig net in a TS of a mycorrhizal root, but the generality of these findings remains unclear. Very few other workers have discovered EcM structures in non-woody (or at least in short-lived) species; whether the Australian species studied by Warcup would have formed fully functional EcM under natural conditions is unknown. However, Fontana (1963), Haselwandter & Read (1981) and several workers since have identified sedges in the genus *Kobresia*

(Cyperaceae) as being EcM. This appears to be the only reliable record of an EcM monocot, if that for *Pandanus* is discounted (B. Moyersoen, unpublished data), and poses interesting problems for a phylogenetic analysis. Numerous reports of EcM plants, especially from field surveys, meet few of the criteria suggested on pp. 266/268, and must be treated with caution. Consequently in the analysis that follows, we have restricted our analysis to the level of the family.

Use of the *rbc*L phylogeny by Soltis *et al.* (1995) offered a novel insight into the evolution of N-fixing symbioses. We have therefore used the same phylogeny to investigate the distribution of ectomycorrhizas, using data obtained from a wide literature, although with a considerable contribution from the dataset compiled by Newman & Reddell (1987), and kindly lent by Dr E. I. Newman. On this basis it appears that the EcM symbiosis (or the propensity to form it) has probably evolved at least twice (Figure 14.2), once or more in the branches leading to the Pinaceae and Gnetales, and once in the root of the clade that includes all the Rosid and Asterid lineages of Chase *et al.* (1993), and possibly the Hamamelids as well. A complication of this analysis is the small number of records of EcM in the Pteropsida. Although some of the records may be discounted, a few seem secure, even though ferns and their allies are largely AM. Did the ability to form EcM arise, therefore, at a very early stage in land plant evolution, becoming lost subsequently in many clades? It seems more plausible that odd EcM ferns are secondary developments, just as we assume the EcM *Kobresia* to be.

Mapping the distribution of EcM symbioses on to the trees of Doyle & Donoghue (1993) allows a comparison of the evolution of woodiness and the EcM habit. They propose two trees of the base of the angiosperm phylogeny, alternatively rooted in the palaeoherb or magnolioid clade. If the magnolioid root is correct, which is the traditional view, then woodiness evolved long before the EcM habit, and many early branches of the tree were woody but not EcM (Figure 14.3*a*). If palaeoherb-rooting is preferred, as is indicated by RNA data (Doyle & Donoghue 1993), then early angiosperms were non-woody (possibly aquatic), but again it is likely that woodiness evolved independently of the ability to form EcM (Figure 14.3*b*).

Among modern plants, the great majority of EcM species are woody, but many woody species are not EcM. Read (1991) has put forward a persuasive hypothesis that the type of mycorrhizal association found is determined by habitat type and the distinct function of the various symbioses. Cool, wet habitats, where nutrient cycling is very slow, favour the ericoid association, in which the fungus can actively mineralize both N and P from organic matter

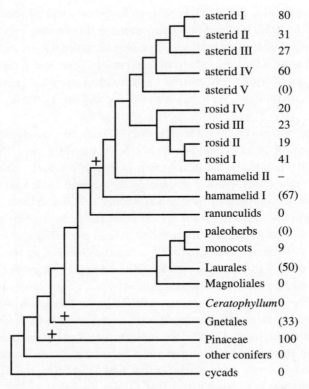

Figure 14.2. Seed-plant phylogeny based on *rbc*L (Chase *et al.* 1993), showing the distribution of ectomycorrhizas. Numbers for EcM represent the percentage of families within each group that contain EcM species; figures in parenthesis are where there are fewer than five families for which data are available and a dash represents 'no data'.

as well as protect the plant from toxic ions. In more productive boreal forests, where nutrient cycling is still potentially limiting, ectomycorrhizal symbioses occur, with mineralization capacities greater than those of AM associations but less effective than ericoid fungi. In the most productive habitats, where N limitation is least and possibly P limitation greater, the ubiquitous AM association is found. The phylogenetic analysis above demonstrates that within the main eudicot lineage, both EcM and AM potentialities are present, and it is safe to make such comparisons as those implied by the hypothesis of Read (1991).

Nevertheless, within tropical forests, which are predominantly AM communities, EcM trees do occur, notably the dipterocarps and some tropical

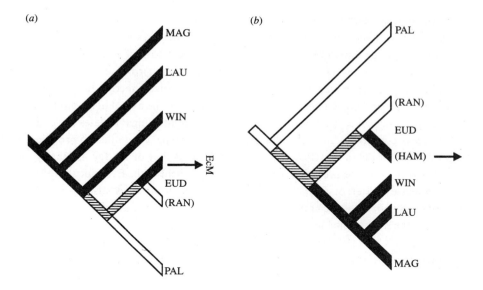

Figure 14.3. Simplified phylogenetic trees of the angiosperms (from Donoghue & Doyle 1993), showing (*a*) magnolioid-rooting, suggesting that woodiness is ancestral, and (*b*) palaeoherb rooting, suggesting that woodiness is derived. In each case the arrow represents the origin of the clade in which virtually all EcM angiosperms (except possibly *Sassafras*, Laurales) occur. Ectomycorrhizality evolved within a woody clade, but not in all woody clades. MAG Magnoliales; WIN Winterales; LAU Laurales; EUD Eudicots; RAN Ranunculids; PAL Palaeoherbs. Solid lines represent woody clades.

legumes. There is little evidence that these EcM species grow on significantly less fertile soils and, although pure EcM stands are sometimes found, EcM species typically occur in mixed communities with AM species (Alexander 1989; Moyersoen 1993). It is possible that these forests are comparable (in terms of nutrient supply and consequent mycorrhizal relationships) to temperate broadleaf forests, where similar AM/EcM mixtures occur.

4.2 Non-mycorrhizal species

In many ways, the most interesting group are those that have lost the ability to form mycorrhizas at all (or at least do not express it). It has been suggested that about 10% of all plant species are normally non-mycorrhizal, including most members of the families Caryophyllaceae, Brassicaceae, Chenopodiaceae, Cyperaceae and Juncaceae. This set of families does not have many obvious ecological characteristics in common. A severe problem exists, however, in defining non-mycorrhizality. A single record of a plant as being non-mycorrhizal is obviously insufficient, let alone a record of some roots of a plant as being so, which is the more likely nature of such a record. Ideally, the criteria on pp. 266/268 need to be applied, at least to the extent of attempting to synthesize mycorrhizas with a range of appropriate fungal isolates under a range of conditions. There are few datasets that can be relied upon to provide this level of detail, but Peat & Fitter (1993) used that of Harley & Harley (1987) and the interpretation of it and other data held in the Ecological Flora Database (Fitter & Peat 1994) for the British flora. They also used an explicitly phylogenetic approach, based on a hierarchical analysis of variance using the taxonomic hierarchy as a surrogate for a true phylogeny.

Peat & Fitter (1993) showed that the majority (80%) of the variance in the proportion of species per genus that are arbuscular mycorrhizal occurs at the level of genus and family. They undertook analyses to compare the ecological characteristics of species that were normally, occasionally, rarely or never arbuscular mycorrhizal, using either species or genera as the units of analysis as appropriate, depending on the results of hierarchical analyses of variance for each characteristic. Non-mycorrhizal species emerged as having thinner roots and smaller seeds than mycorrhizal species, and also as occurring in habitats of less extreme pH. They did not, however, have any special relationship with soil fertility or moisture status. The most striking result was that non-mycorrhizal perennial species occur in fewer habitat types on average than do mycorrhizal perennials; in other words, they are habitat specialists, and this specialization is not simply due to the fact that some non-mycorrhizal species favour particular habitat types, such as disturbed ground or wetlands, as the relationship persists when wetland species and halophytes are omitted, and is not stronger when only annuals are considered.

Nevertheless, it appears that non-mycorrhizality has evolved most frequently in two distinct groups of species. The first group is of species characteristic of disturbed habitats where a mycorrhizal network is unlikely to develop, retarding the development of colonization by the fungus of newly extending root systems, and so forcing plants to become dependent on their own resource acquisition abilities for at least part of their life cycle. In these habitats, competition for nutrients is unlikely to be very important, because low root length densities in soil lead to a low probability that depletion zones around roots will overlap. Second, non-mycorrhizality has evolved in species of wet habitats where diffusion of poorly mobile ions is faster than in dry soils and where the benefits of mycorrhizality, at least in terms of phosphate uptake, are smaller. Non-mycorrhizality is therefore a specialist trait, occurring in relatively few species and restricting their ecological amplitude.

Consequently, one can regard AM species as generalists, with all other types (non-mycorrhizal, EcM, ericoid, etc.) as specialists in one way or another; this view is consistent with that of Read (1991). The curiosity of the AM symbiosis is that it appears to have evolved as a direct response to the inability of early, rootless land plants to acquire phosphate (Pyrozinski & Malloch, 1975). However, complex root systems have been a feature of land plants for a long time, although the fossil record is surprisingly unclear as to exactly how long. Many modern plants that have intensely branched root systems, such as the Poaceae, are normally AM when growing under natural conditions. Fitter & Merryweather (1992) demonstrated that there is no apparent benefit to the plant from the association in these circumstances, if benefit is to be defined in terms of increased P acquisition. This is even more true of plants that possess 'hairy root clusters' (Lamont 1993), dense groupings of lateral roots that are highly effective at absorbing nutrients from infertile soils, and that develop frequently in non-mycorrhizal taxa such as the Proteaceae.

The question therefore arises as to whether mycorrhizal associations in plants with well-developed root systems are non-mutualistic, perhaps even parasitic, or whether there are other benefits to the plant. Newsham *et al.* (1995*a*) have conclusively demonstrated that in the grass *Vulpia ciliata*, which is normally an AM species, there is no enhancement in P uptake as a consequence of colonization, but that the mycorrhizal fungus does protect the plant against other fungi (notably *Fusarium oxysporum*) that are root pathogens. They postulated (Newsham *et al.* 1995*b*) that as root systems evolved more complex branching patterns, and so became self-sufficient in terms of P acquisition in most soils, they also became more vulnerable to pathogens, as many such fungi attack roots where lateral roots penetrate the cortex. The ability to protect plants against pathogens therefore maintained the AM association as a mutualism.

5 Conclusions

Although root-microbe symbioses are ubiquitous among land plants and play a key role in their ecology, only one dataset has proved amenable to a full comparative analysis: that of Peat & Fitter (1993). This showed that non-mycorrhizal species do have consistent characteristics, of which the most notable is that they are habitat specialists. It is likely that similar insights could be obtained from a comparative analysis of ectomycorrhizal species were an appropriate dataset available.

Our ability to undertake systematic comparative analyses of root–microbe symbioses of plants is limited both by the rudimentary state of our knowledge of plant phylogeny and by the inadequacy of the data about the symbioses themselves, which are only really well known within a few tightly defined groups (such as the rhizobium nodule symbiosis) that do not lend themselves to comparative analysis. The new phylogenies that are appearing have already been used to demonstrate that N-fixing root symbioses all occur in a single clade (Soltis et al. 1995), and in this chapter we demonstrate that the same phylogeny suggests that ectomycorrhizal association appeared at least twice, in the Pinaceae and in the eudicots. Future comparisons of ectomycorrhizal and other plant species must take this fundamental distinction into account.

We are grateful to E. I. Newman for making available the original data used by Newman & Reddell (1987), to J. M. Trappe for information and advice, and to I. J. Alexander, M. H. Williamson, J. P. W. Young and D. J. Read for valuable comments on the manuscript.

References

Alexander, I. J. (1989). Systematics and ecology of ectomycorrhizal legumes. *Monogr. Syst. Missouri Botanical Garden* **29**, 607–624.

Alexander, C. & Hadley, G. (1984). The effect of mycorrhizal infection of *Goodyera repens* and its control by fungicide. *New Phytologist* **97**, 391–400.

Cázares, E. & Smith, J. E. (1996). Occurrence of vesicular-arbuscular mycorrhizae in *Pseudotsuga menziesii* and *Tsuga heterophylla* seedlings grown in Oregon Coast Range soils. *Mycorrhiza* **6**, 65–67.

Cázares, E. & Trappe, J. M. (1993). Vesicular endophytes in roots of the Pinaceae. *Mycorrhiza* **2**, 153–156.

Chase, M. W. & 41 others (1993). Phylogenetics of seed plants: an analysis of nucleotide sequences from the plastid gene *rbc*L. *Annals of the Missouri Botanical Garden* **80**, 528–580.

Cournoyer, B., Gouy, M. & Normand, P. (1993). Molecular phylogeny of the symbiotic actinomycetes of the genus *Frankia* matches host-plant infection processes. *Molecular Biology and Evolution* **10**, 1303–1316.

Cronquist, A. (1981). *An integrated system of classification of flowering plants.* New York: Columbia University Press.

Cullings, K. (1994). Molecular phylogeny of the Monotropoideae (Ericaceae) with a note on the placement of the Pyroloideae. *Journal of Evolutionary Biology* **7**, 501–516.

Cullings, K., Szaro, T. M. & Bruns, T. D. (1996). Evolution of extreme specialization within a lineage of ectomycorrhizal epiparasites. *Nature* **379**, 63–66.

Doyle, J. J. (1994). Phylogeny of the legume family: an approach to understanding the origins of nodulation. *Annual Review of Ecology and Systematics* **25**, 325–349.

Doyle, J. A. & Donoghue, M. J. (1993). Phylogenies and angiosperm diversification. *Paleobiology* **19**, 141–167.

Duckett, J. G. & Read, D. J. (1995). Ericoid mycorrhizas and rhizoid-ascomycete associations in liverworts share the same mycrosymbiont: isolation of the partners and resynthesis of the associations *in vitro*. *New Phytologist* **129**, 439–447.

Field, C. & Mooney, H. A. (1986). The photosynthesis-nitrogen relationship in wild plants. *In On the economy of plant form and function* (ed. T. J. Givnish), pp. 25–55. Cambridge University Press.

Fitter, A. H. & Peat, H. J. (1994). The Ecological Flora Database. *Journal of Ecology* **82**, 415–425.

Fitter, A. H. & Merryweather, J. W. (1992). Why are some plants more mycorrhizal than others? An ecological enquiry. In *Mycorrhizas in Ecosystems* (ed. D. J. Read, I. J. Alexander, A. H. Fitter & D. H. Lewis), pp. 26–36. Wallingford: CABI.

Fontana, A. (1963). Micorriza ectotrofiche in una ciperacea: *Kobresia bellardii*. *Degl Giornale Botanico Italiano* **70**, 639–641.

Gianinazzi-Pearson, V., Dumas-Gaudot, E, Gollotte, A., Tahiri-Alaoui, A. & Gianinazzi, S. (1996). Cellular and molecular defence-related root responses to invasion by arbuscular mycorrhizal fungi. *New Phytologist* **133**, 45–57.

Hall, I. R. (1976). Vesicular mycorrhiza infection in the orchid *Corybas macranthus*. *Transactions of the British Mycological Society* **66**, 160.

Harley, J. L. & Harley, E. L. (1987). A check-list of mycorrhiza in the British flora. *New Phytologist* **105**, 1–102.

Harley, J. L. & Smith, S. E. (1983). *Mycorrhizal symbiosis*. London: Academic Press.

Haselwandter, K. & Read, D. J. (1981). Observations on the mycorrhizal status of some alpine plant communities. *New Phytologist* **88**, 341–352.

Kidston, R. & Lang, W. H. (1921). Old Red Sandstone plants showing structure, from the Rhynie Chert bed, Aberdeenshire. Part 5. *Transactions of the Royal Society of Edinburgh* **52**, 855–902.

Lamont, B. (1993). Why are hairy root clusters so abundant in the most nutrient-impoverished soils of Australia? *Plant and Soil* **155/156**, 269–272.

Leake, J. R. (1994). The biology of mycoheterotrophic (saprophytic) plants. *New Phytologist* **127**, 171–216.

McKey, D. (1994). Legumes and nitrogen: the evolutionary ecology of a nitrogen-demanding lifestyle. *Advances in Legume Systematics* **5**, 211–228.

Moyersoen, B. (1993). *Ectomicorrizas y micorrizas vesciculo-arbusculares en Caatinga Amazonica del Sur de Venezuela.* Scientia Guaianae, 3. Huber Caracas.

Newman, E. I. & Reddell, P. (1987). The distribution of mycorrhizas among families of vascular plants. *New Phytologist* 106, 745–751.

Newsham, K. K., Fitter, A. H. & Watkinson, A. R. (1995a). Arbuscular mycorrhiza protect an annual grass from root pathogenic fungi in the field. *Journal of Ecology* 83, 991–1000.

Newsham, K. K., Fitter, A. H. & Watkinson, A. R. (1995b). Multi-functionality and biodiversity in arbuscular mycorrhizas. *Trends in Ecology and Evolution* 10, 407–411.

O'Dell, T. E., Massicotte, H. B. & Trappe, J. M. (1993). Root colonization of *Lupinus latifolius* Agardh. and *Pinus contorta* Dougl. by *Phialocephala fortinii* Wang & Wilcox. *New Phytologist* 124, 93–100.

Peat, H. J. & Fitter, A. H. (1993). The distribution of arbuscular mycorrhizas in the British Flora. *New Phytologist* 125, 843–854.

Pirozynski, K. A. & Malloch, D. W. (1975). The origin of land plants: a matter of mycotrophism. *BioSystems* 6, 153–164.

Read, D. J. (1991). Mycorrhizas in ecosystems. *Experientia* 47, 376–391.

Simon, L, Bousquet, R. C., Levesque, C. & Lalonde, M. (1993). Origin and diversification of endomycorrhizal fungi and coincidence with vascular land plants. *Nature* 363, 67–69.

Soltis, D. E., Soltis, P. S., Morgan, D. R. *et al.* (1995). Chloroplast gene sequence data suggest a single origin of the predisposition for symbiotic nitrogen fixation in angiosperms. *Proceedings of the National Academy of Sciences of the U.S.A.* 92, 2647–2651.

Stoyke, G., Egger, K. N. & Currah, R. S. (1992). Characterisation of sterile, endophytic fungi from the mycorrhizae of sub-alpine plants. *Canadian Journal of Botany* 70, 2009–2011.

Taylor, T. N., Remy, W., Hass, H. & Kerp, H. (1995). Fossil arbuscular mycorrhizae from the early Devonian. *Mycologia* 87, 560–573.

Trappe, J. M. (1987). Phylogenetic and ecological aspects of mycotrophy in the angiosperms from an evolutionary standpoint. In *Ecophysiology of VA mycorrhizal plants* (ed. G. Safir), pp. 5–25. Boca Raton: CRC Press.

Warcup, J. H. (1980). Ectomycorrhizal associations of Australian indigenous plants. *New Phytologist* 85, 531–535.

Warcup, J. H. & McGee, P. A. (1983). The mycorrhizal associations of some Australian Asteraceae. *New Phytologist* 95, 667–672.

Young, J. P. W. & Haukka, K. E. (1996). Diversity and phylogeny of rhizobia. *New Phytologist* 133, 87–94.

15 · Competitive ability: definitions, contingency and correlated traits

Deborah E. Goldberg

1 Introduction

Ecologists generally agree that competition often has a major role in structuring communities, especially in plants (Begon *et al.* 1990; Crawley 1990; Goldberg & Barton 1992). However, there is much less agreement about the two necessary components of any comprehensive model to explore this role: patterns in variation in the importance of the process of competition and in the traits that lead to success, given that competition is an important process. In this chapter, I focus on the latter component and specifically, how to evaluate relationships between individual plant traits and competitive success.

Three major obstacles have stymied attempts to examine relationships between traits and competitive ability. First, competitive ability can be a rather vague concept that is defined in very different ways by different authors (Milne 1961; Abrams 1987; Goldberg 1990; Grace 1990). Therefore, apparent contrasts or similarities between predictions from different bodies of theory or between theory and empirical observation are not necessarily valid. Second, even with consistent definitions of competitive ability, it is potentially a highly contingent trait, which could greatly restrict the domains in which consistent correlations between traits and competitive ability can be found. The third obstacle is logistical; competitive ability can be a very time-consuming property to measure compared to life history, morphological or physiological traits, resulting in an extremely limited database to test relationships. As a contribution toward an eventual synthesis of the relationship between individual plant traits and competitive ability, I devote most of this chapter to 'clearing the decks' in preparation for eventual testing of predictions about these relationships by addressing each of these three obstacles.

2 Definitions of competitive ability

How to define competitive ability has been a source of debate among ecologists for decades (Milne 1961; Abrams 1987; Thompson 1987; Thompson & Grime 1988; Tilman 1987) and, as discussed below, differences in definitions of competitive ability are at least partly responsible for different predictions of traits correlated with competitive ability (Goldberg 1990; Grace 1990). The key problem here is that much of formal theory is based on populations at equilibrium but the vast majority of experimental work on competition has been at the individual level and short term. Despite the many recent examples of changes in outcome of experiments over time (e.g. Tilman & Wedin 1991a; Grace et al. 1992; Heske et al. 1994; Inouye & Tilman 1995), logistical and temporal constraints will mean that long-term experiments are simply not possible except in a few model systems. The majority of experimental work on competition will probably continue to be individual level and short term. The value of such work in agronomy and forestry is apparent. However, if it is also to be profitable in understanding patterns in natural communities, it is incumbent on ecologists to (a) formulate reasonable individual-level definitions that are comparable among species and studies, and (b) investigate how individual level measures of competitive ability relate to long-term population-level outcomes of competitive interactions.

2.1 Individual-level definitions of competitive ability

The magnitude of competition at the individual level can be quantified as the per unit effect of individuals of some neighbour taxon on the response of some target taxon, where 'unit' can refer to individuals, biomass or any other measure of abundance, and response is measured as some component of fitness of target individuals at different abundances of the neighbours. While this single value expresses competitive ability of both taxa in a particular interacting pair, comparisons of this value among taxa, i.e. comparisons of competitive ability, can be done in two distinct ways that have very different interpretations (Jacquard 1968; Goldberg & Werner 1983). Comparisons among neighbour species assess ability to suppress other plants or competitive effect, while comparisons among target species assess resistance to suppression or competitive response. These two measures of competitive ability are not necessarily positively correlated (e.g. Goldberg & Landa 1991; Keddy et al. 1994) and therefore it is critical that both predictions of the traits correlated with competitive ability and empirical tests of those predictions be explicitly separated.

To compare competitive response, target performance should be standardized to performance in the absence of neighbours to eliminate differences among target taxa due solely to differences in size or growth rate rather than in response to neighbours (Goldberg & Scheiner 1993; Grace 1995). While this seems obvious, its implications are not. Most importantly, it means that, in addition to their other well known problems, replacement series experiments cannot be used to quantify competitive response to other species of neighbours because they only provide a measure of the relative intensity of intra- to interspecific competition. This could vary among target species solely because of variation in intraspecific competition rather than in response to interspecific competition.

2.2 Relating theory to experimental results: predicting traits determining competitive ability

Predicting which traits should be correlated with competitive ability has been one of the major goals of plant community ecology in recent years, with a wide variety of sometimes contrasting traits predicted (Table 15.1). The most comprehensive set of predictions for plants are those of Tilman (1988, 1990a), Smith & Huston (1989), Grime (1977, 1979) and Colosanti & Grime (1993). These authors all make predictions of both short-term and long-term outcomes (although they do not necessarily call patterns at both timescales reflective of competitive ability) and at both high and low productivity where productivity is usually determined by a gradient in soil resources (nutrients and/or water). All are based at least in part on formal theory (usually simulation modelling, necessitated by the inclusion of size structure, which in turn is critical to include for biologically reasonable models in terrestrial plant communities). Table 15.1 also includes some predictions for sets of traits correlated with competitive ability for particular resources or in a more limited range of environments.

Because the models summarized in Table 15.1 differ greatly in what is explicitly defined as competitive ability, their predictions are not directly comparable. For example, Tilman (1987) has defined competitive ability as the ability of a population to dominate at equilibrium and generated predictions about traits correlated with this measure of competitive ability from a simulation model of competition for nutrients and light along a nutrient supply gradient (Tilman 1988, 1990a). Similarly, Smith & Huston (1989) make predictions about competitively superior morphologies for populations at equilibrium over a water supply gradient (see their Figure 15.7), although they consider only competition for light throughout the gradient. However, competitive ability for populations at equilibrium is clearly impractical to measure for many, if not most, plants and therefore testing these predictions about traits correlated with competitive ability will most often require a surrogate. Goldberg (1990) argued that response of individuals in strongly size-uneven situations could be an effective surrogate (see also Wilson & Tilman 1995). That is, the competitive response of seedlings or juveniles to established vegetation should reflect rankings of long-term response of populations at equilibrium, because in a population at equilibrium, seedlings must be able to tolerate the depleted resource levels imposed by surrounding adults. Consistent with this argument, using MacArthur's (1972) consumer-resource equations, J. H. Vandermeer & D. E. Goldberg (unpublished results) show that ability of individuals to deplete resources is irrelevant to the equilibrium outcome of competition for a single resource and only ability to tolerate low levels of the resource determines the outcome.

Both Tilman (1988) and Smith & Huston (1989) also make predictions about traits leading to dominance earlier in succession, which also apply to disturbed sites. Translation to individual-level measures of competitive ability is again problematical and I tentatively suggest that early successional dominance reflects individual-level response when plants are competing at roughly even sizes (Table 15.1). That is, at the initiation of interactions among a group of individuals, resources have not yet been strongly depleted and resource pre-emption by any individual is possible (see also Goldberg 1990).

In contrast, Grime (1977) has defined competitive ability as the ability of individuals to take up resources rapidly and prevent their use by other organisms. This seems to correspond with short-term competitive effect, i.e. ability to suppress other plants (Goldberg 1990). Because Grime argues that competitive hierarchies are consistent between environments but that competition is relatively unimportant in unproductive environments, the predictions in Table 15.1 hold for competitive ability in low and high productivity, but not for the traits of dominant species in low and high productivity.

The assignments of predictions from different sorts of models or components of models to the different ways of measuring individual-level competitive ability in Table 15.1 are, of course, themselves hypotheses that should be tested both theoretically and empirically. To the extent that these assignments are correct, the apparently large differences in prediction among columns in Table 15.1 are not necessarily contradictory because predictions are for different components of competitive ability and should be tested independently.

3 Consistency and contingency of competitive hierarchies

Before attempting to evaluate the predictions in Table 15.1, it is necessary to investigate to what extent competitive ability can be regarded as a characteristic of a particular taxon at all and the domains in which consistent correlations might occur.

The database for this investigation included both a quantitative component (all papers meeting the criteria from *Ecology* and *Journal of Ecology* over the 17-year period from 1979 to 1995) and a less objective and smaller selection of studies, mostly consisting of those cited in papers from the quantitative database as potentially relevant to the questions. I chose *Ecology* and *Journal of Ecology* for the quantitative survey because they included over 60% of the experiments found in an earlier survey of seven ecological journals of field competition experiments in plants for the first ten years of this period (Goldberg & Barton 1992). This approach undoubtedly excludes a large number of relevant studies, especially in agronomic journals, but is likely to be reasonably representative for plants in naturally occurring communities. To enable comparisons of competitive response as well as effect and because of their other limitations, substitutive experiments conducted at a single density were excluded. With this exception, all experiments on interspecific interactions between living plants (i.e. purely litter effects were excluded) that met at least one of the following three criteria were included: (1) a minimum of two neighbour and two target taxa for testing contingency of hierarchies among species, (2) a minimum of two environments (experimental or natural) and two neighbour or two target taxa for testing contingency of effect or response hierarchies among environments, or (3) a minimum of three targets or three neighbours and quantitative estimation of at least one trait to be related to competitive ability. Taxa were usually species but sometimes genotypes within species and sometimes groups of species. The studies and data extracted from them are listed in Table 15.1.

3.1 Consistency vs contingency of competitive hierarchies among species

The whole notion of testing correlations between traits and competitive ability assumes that competitive ability is a property of a particular taxon and

Table 15.1. *Summary of predictions about traits correlated with competitive ability*

('high' (or 'low') indicates that competitive ability increases (or decreases) with increasing values of the trait. All predictions are for a measure of individual-level competitive ability as described in the text and not dominance *per se* to avoid confounding these predictions with the importance of the process of competition.)

Trait	Response size-uneven (low productivity)	Response size-uneven (high productivity)	Effect size-even or uneven (all environments?)	Response size-even (all environments?)
Allocation to roots	high[a,b]	low[a]		low[a]
Allocation to leaves	low[a,b]	low[a]		low[a]
Allocation to stems	low[a,b]	high[a]		low[a]
Allocation to above ground	low[a,b,d]	high[a,d]		high[a]
Allocation to reproductive output	low[a,b]	low[a]	low[c]	high[a]
Plant or leaf height	low[a,b,d]	high[a,d]		low[a,g]
maxRGR	low[a,b,d]	low[a,d]		high[a,d]
Leaf size			high[c]	
Biomass/plant			high[d]	
Specific root length (length/mass)			high[c]	
Leaf area ratio (lf area/plant mass)			high[f]	
Lateral spread			high[c]	
Plant longevity	low[d]	high[d]	high[c]	
Tissue N, N productivity	low[b,e]	high[e]		
Tissue longevity, nutrient retention	high[b,e]	low[e]	low[e]	
Defence investment	high[b]			
Maximal nutrient uptake/mass	low[b]		high[c]	

Efficiency nutrient uptake	high[b]		high[c]	
Litter production				
Shade tolerance	low[b,d]	high[a,d]	low[c]	low[a,d]
Drought tolerance	high[d]	low[d]	low[c]	low[d]
Low nutrient tolerance	high[b,e]	low[a,e]	low[c]	low[a,e]

[a] From Tilman 1988.
[b] From Tilman 1990b.
[c] From Grime 1977.
[d] From Smith & Huston 1989.
[e] From Berendse & Elberse 1990.
[f] From Caldwell & Richards 1986.
[g] From Givnish 1982.

not of a particular *combination* of taxa. If this assumption is correct, rankings of competitive effects of neighbours would be similar among targets, and/or rankings of competitive responses of targets would be similar among neighbours, i.e. both effect and response competitive hierarchies would be transitive within a given environment. The assumption is most likely to be correct if all the taxa involved in a comparison are competing for the same resources, which a number of authors have argued is more likely to be true for plants (and sessile animals) than for other organisms (Goldberg & Werner 1983; Shmida & Ellner 1984; Hubbell & Foster 1986; Mahdi *et al.* 1989).

Keddy & Shipley (1989) and Shipley (1993) tested this assumption and found generally consistent hierarchies; however their database and analytical approach have two limitations. First, they relied on matrices of substitutive experiments, for the good reason that these were the main source of available data. Herben & Krahulec (1990) and Silvertown & Dale (1991) have aptly critiqued the use of substitutive experiments in this context because of the sensitivity of their results to density and to plant size (but see Shipley & Keddy 1994). In addition, as already noted, it is impossible to compare competitive responses with this design. Second, the analytical approach they use does not separate consistency of competitive effect and of competitive response hierarchies; it is entirely possible in principle that one but not the other is consistent.

The database included 21 experiments that could be used to test the consistency of either effect and/or response hierachies among species within a single environment (Table 15.2*a*). This excludes many studies that appeared to have gathered such data but the results were not analysed to test directly for transitivity or were not presented in such a way that I could reconstruct the competitive matrix. Two statistical approaches were used. Where ANOVAS testing explicitly for the appropriate interaction term (see Goldberg & Scheiner 1993) were provided, the absence of a significant interaction term was taken as evidence for consistency. Where a competition matrix was provided or could be reconstructed easily, I used Kendall's coefficient of concordance to test for consistency of rankings of competitive effect of neighbours among different targets or of competitive response of targets among different neighbours. Significant concordance was taken as evidence of consistency. Thus, in neither case was an absolute criterion of complete transitivity used. In addition, for some studies in which neither ANOVAS nor concordance were provided or could be calculated, I used the author's interpretations or my own inspection of the data to infer consistency or contingency – these studies are counted separately in Table 15.2.

Table 15.2. *Numbers of studies showing consistency and contingency of competitive effect and response among different competitors within a single environment (a) and among environments for a given competitor (b)*

(Individual studies and their results are listed in Appendix I. Values in parentheses are the number of studies with conclusions on consistency or contingency confirmed by statistical analyses while values outside of parentheses include studies where the conclusions are reached by data inspection. 'Variable' results indicate that more than one test could be conducted within a study and results differ among tests (e.g. more than one environment for consistency with respect to identity of competitors or more than one type of environment for consistency with respect to environments).)

	consistent	contingent	variable
(a) *Identity of competitor*			
Effect	14 (13)	6 (6)	1 (1)
Response	12 (10)	1 (1)	2 (2)
(b) *Environment*			
Effect	6 (5)	7 (6)	1 (1)
Response	11 (7)	9 (6)	2 (2)

Hierarchies are almost entirely consistent for competitive response and largely consistent (approximately 2:1) for competitive effect (Table 15.2a). This contrast between effect and response is the exact opposite of that predicted by Goldberg (1990) on the basis of traits predicted to be correlated with effect and response. One possible explanation is that competitive responses are more similar among species and therefore analysis of changes in rankings is meaningless because species with different ranks are identical in statistical terms. This seems to be the case in the matrix analysed by Goldberg & Landa (1991), but it is clear in some other cases that rankings are still consistent even when differences in magnitude of response are large.

Potentially important caveats for this analysis are that the majority of the 22 experiments with appropriate data were conducted in glasshouses or common gardens (16), were short-term (16 less than 1 year), and were for even-sized interactions among seedlings (17) (Appendix I). However, the limited data available do not show any consistent relationships between these variables and the consistency or contingency of hierarchies (Appendix I).

Thus, the search for correlations of traits with competitive ability in plants seems reasonable, especially for response.

3.2 Consistency and contingency of competitive hierarchies among environments

Unlike the case of consistent hierarchies among species, consistency of hierarchies among environments is not a necessary assumption for the entire approach of looking for correlations of traits with competitive ability. But it is critical for deciding the level at which to look for such correlations and how complicated or general any correlation structure will be. In addition, whether or not competitive ability is consistent among environments is at the centre of a major controversy in plant ecology. Grime (1977) has argued for a 'unified concept of competitive ability', i.e. positive correlations in competitive ability for different resources and therefore in different environments when different resources are likely to be limiting, while Tilman (1988, 1990b) has argued that trade-offs in competitive ability for different resources is a fundamental principle underlying patterns of plant distribution. These two positions may not be as different as they sound if the earlier arguments are correct that Tilman's definition of competitive ability corresponds to competitive response in size-uneven situations, and that Grime's definition of competitive ability corresponds to competitive effect. Consistency of competitive effect hierarchies but contingency of competitive response hierarchies among environments could be viewed as consistent with both sides of this divisive issue (Goldberg 1990).

The database includes 30 studies that allow comparisons of competitive effect and/or response hierarchies among environments, and no clear answer emerges for either one (Table 15.2b). About half of experiments testing for both competitive effect and competitive response were consistent between environments and about half were contingent. Again, many of the available experiments were in a glasshouse or common garden (11), short-term (10 less than 1 year) and involved only seedling-seedling interactions (14), but the limited data available do not suggest any patterns in consistency vs contingency with respect to these variables (Appendix I).

Clearly, there will be no simple answer to whether competitive ability is consistent among environments or hierarchies change, and there is currently neither sufficient theory nor empirical work to guide us as to when hierarchies are likely to be consistent and when not. However, sufficient data do exist that both effect and response hierarchies can differ between environments that it is probably unreasonable to assume *a priori* consistency in any particular case.

4 Correlations of traits with competitive ability

A minimum of three species is needed to even begin to relate traits to competitive ability and this is clearly statistically insufficient to establish relationships. Unless species are carefully chosen, even statistically reasonable numbers of taxa will often be insufficient because of possible phylogenetic effects. In the context of understanding the consequences of traits, phylogenetic relationships may be important if apparent relationships between a trait and competitive ability are actually due to some other, shared trait that has not been measured. The distribution of numbers of targets or neighbours in the database is strongly skewed towards very few taxa with a median value between three and four taxa for both response and effect (Appendix I). This is despite a strong bias towards inclusion of larger numbers of taxa because of the criteria used (see above). Further, most of those with three or more taxa do not also include explicit relationships with traits and those few that do only use a single environment, most often a relatively productive one (Gross 1984; Goldberg 1987a; Gaudet & Keddy 1988; Popma & Bongers 1988; Goldberg & Landa 1991; Reader 1993). Finally, the set of studies listed above that have three or more species and trait information do not, in any case, incorporate phylogenetic effects in the analyses, making it impossible to separate if the similarities in competitive ability are because of the shared trait values or because of other traits that might also be in common due to common ancestry.

Thus, the database to test the predictions in Table 15.1 in any kind of rigorous and general way is just about non-existent. This is not to say that excellent data on the traits determining competitive ability do not exist for particular ecological systems. Much is known about the mechanisms of competitive interaction from a few exceptionally detailed research programmes on specific systems that cannot be reviewed here for lack of space (e.g. Eissenstat & Caldwell 1987, 1988, 1989; Aerts *et al.* 1990, 1991; Berendse & Elberse 1990; Tilman 1990a; Tilman & Wedin 1991a, b; Wedin & Tilman 1993). Nevertheless, the ability to generalize from this detailed knowledge will also require studies that cover much broader ranges of species, albeit with the cost of much less mechanistic detail.

5 Expanding the dataset: problems and possible solutions

The lack of large numbers of species within studies on competitive ability is not because plant ecologists have not recognized the importance of doing so but because the logistics of such experiments in even a single simple glasshouse environment are nightmarish, as the number of necessary experiments go up exponentially with the number of species for a complete matrix of pairwise interactions. Therefore, expanding the sample sizes for analysing relationships between traits and the outcome of interactions and rankings of competitive ability is going to require making some assumptions to reduce the dimensionality of the systems. One such simplifying assumption is to study competitive response to diffuse competition from all vegetation, or at most, particular growth forms, reducing a huge matrix of possible interactions to a single column vector of multiple target species with a single neighbour 'taxon'. This is, in fact, the most common approach to studying competitive interactions in the field (Goldberg & Barton 1992), which may reflect the intuition of biologists that the simplifying assumptions it entails are reasonable. Specifically, this approach assumes that relative competitive response is more important than effect to persistence and abundance of a taxon within a community, and that competitive effects are equivalent among neighbour taxa. The first assumption was discussed earlier but clearly needs to be tested explicitly. While the second assumption is clearly not correct under highly controlled conditions, it may not be too awful on a per unit size basis under field conditions (Goldberg 1996). It may also be reasonable to lump neighbours if effects are not equivalent but rankings of neighbours are consistent among targets, which was true in approximately two-thirds of the cases in the quantitative survey (Table 15.2a).

Within this set of simplifying assumptions, at least two primarily field-oriented approaches are possible to garner data on competitive ability on large numbers of species within a single study. The more conventional approach is to compare response of individuals in the presence *vs* absence of existing vegetation. The advantage of this approach is that it is directly 'field-relevant' to the particular system under study and is relatively simple in principle, if time consuming to actually carry out. There are, however, at least two disadvantages. First, it is best to use separate plots for each target species to avoid interactions among targets that could confound the results, requiring a linear (but at least not exponential) increase in number of plots with number of target species. Second, because only a single abundance of the neighbour (total vegetation or growth forms) is used, extrapolation to other sites or times with different neighbour abundances is limited, given the typical non-linearity of competitive interactions in plants.

An alternative approach to studying competitive ability for large numbers of species provides potential solutions to both these limitations. Goldberg *et al.* (1995) suggested a design called the community density series that is a simple extension of the classic yield-density experiment in agronomy. If density of the total community is varied, while holding initial relative abundances constant, subsequent changes in relative abundances along the community density gradient should reflect effects of plant–plant interactions. Specifically, the slope of a regression of eventual relative abundance on initial community density for a particular species is a measure of its community-context competitive ability. A single experiment thus yields estimates of competitive ability for all the species in a community over a gradient in density from much below to above naturally occurring abundances. Preliminary results using relatively simple communities of annual plants on stabilized sand dunes suggest that the method is feasible (D. E. Goldberg, R. Turkington and L. Olsvig-Whittaker, unpublished data).

6 Conclusions

For ecologists, the primary motivation behind understanding the links between individual plant traits and competitive ability is as a component of a larger program that links competitive ability to relative abundances and dynamics in natural communities. Two broad types of approaches to this general program can be caricatured as follows. One extreme is primarily experimental with detailed and long-term study of the mechanisms of interaction among few species. The other extreme is observational and based on broad surveys of large numbers of species with relatively easily measured traits, often garnered from the literature or from floras.

Both of these approaches provide important and useful knowledge that cannot be gained otherwise. However, the main message of this chapter is that an intermediate approach that links these two is equally important but currently largely missing: relatively short-term experiments that provide a link of inference missing in purely observational studies but that are designed to be applied to large numbers of species, so that results are rigorously generalizable. Westoby *et al.* (this volume) have provided an outstanding example of such an integration focusing on the consequences of seed size in terms of numerous processes, including competition. Keddy and his colleagues have pioneered this approach with respect to competitive ability, but so far largely applied it narrowly to only competitive effect among approximately equal-sized plants in productive environments (e.g. Gaudet & Keddy 1988, 1995; but see Keddy *et al.* 1994).

Applying this intermediate approach in a more general way is critical for broad testing of predictions about the traits determining ability to compete in natural vegetation and, from there, the relationship of competitive ability to abundance and dynamics in natural vegetation. Ecologists are increasingly being asked to make predictions about community dynamics in the face of anthropogenic environmental change. If more than a few species and systems are to be investigated to serve as a basis for assessing general models of community dynamics, it is essential that some way be found to scale up from short-term individual-level interactions to long-term population-level outcomes of competition and to circumvent the logistical constraints on studying competitive interactions in highly complex, diverse communities. The suggestions in this chapter are meant to provoke discussion of such issues.

The ideas presented in this chapter are based on research funded by the National Science Foundation, the Binational (US-Israel) Science Foundation and the Office of the Vice President for Research of the University of Michigan; I am grateful for their support. I also thank Drew Barton and Chad Hershock for help with the literature survey and Katie Nash for suggesting the terms size-even and uneven.

References

Aarssen, L. W. (1988). 'Pecking order' of four plant species from pastures of different ages. *Oikos* **51**, 3–12.

Abrams, P. A. (1987). On classifying interactions between populations. *Oecologia* **73**, 272–281.

Aerts, R., Berendse, F., deCaluwe, H. & Schmitz, M. (1990). Competition in heathland along an experimental gradient of nutrient availability. *Oikos* **57**, 310–318.

Aerts, R., Boot, R. G. A. & van der Aart, P. J. M. (1991). The relation between above- and belowground biomass allocation patterns and competitive ability. *Oecologia* **87**, 551–559.

Austin, M. P. & Austin, B. O. (1980). Behaviour of experimental plant communities along a nutrient gradient. *Journal of Ecology* **68**, 891–918.

Begon, M., Harper, J. L. & Townsend, C. R. (1990). *Ecology*. Oxford: Blackwell Scientific Publications.

Berendse, F. & Elberse, W. T. (1990). Competition, succession, and nutrient availability. In *Perspectives in plant competition* (ed. J. Grace & D. Tilman), pp. 93–116. San Diego: Academic Press.

Berkowitz, A. R., Canham, C. D. & Kelly, V. R. (1995). Competition vs. facilitation of tree seedling growth and survival in early successional communities. *Ecology* **76**, 1156–1168.

Bertness, M. D. & Yeh, S. M. (1994). Cooperative and competitive interactions in the recruitment of marsh elders. *Ecology* **75**, 2416–2429.

Caldwell, M. M. & Richards, J. H. (1986). Competing root systems: morphology and models of absorption. In *On the economy of plant form and function* (ed. T. J. Givnish), pp. 251–273. Cambridge: Cambridge University Press.

Campbell, B. D. & Grime, J. P. (1992). An experimental test of plant strategy theory. *Ecology* **73**, 15–29.

Colasanti, R. L. & Grime, J. P. (1993). Resource dynamics and vegetation processes: a deterministic model using two-dimensional cellular automata. *Functional Ecology* **7**, 169–176.

Crawley, M. J. (1990). The population dynamics of plants. *Philosophical Transactions of the Royal Society of London* **B 330**, 125–140.

De Steven, D. (1991*a*). Experiments on mechanisms of tree establishment in old-field succession: seedling emergence. *Ecology* **72**, 1066–1075.

De Steven, D. (1991*b*). Experiments on mechanisms of tree establishment in old-field succession: seedling survival and growth. *Ecology* **72**, 1076–1088.

Eissenstat, D. M. & Caldwell, M. M. (1987). Characteristics of successful competitors: An evaluation of potential growth rate in two cold desert tussock grasses. *Oecologia* **71**, 167–173.

Eissenstat, D. M. & Caldwell, M. M. (1988). Competitive ability is linked to rates of water extraction: A field study of two aridland tussock grasses. *Oecologia* **75**, 1–7.

Eissenstat, D. M. & Caldwell, M. M. (1989). Invasive root growth into disturbed soil of two tussock grasses that differ in competitive effectiveness. *Funtional Ecology* **3**, 345–353.

Fowler, N. L. (1990). The effects of competition and environmental heterogeneity on three coexisting grasses. *Journal of Ecology* **78**, 389–402.

Gaudet, C. L. & Keddy, P. A. (1988). A comparative approach to predicting competitive ability from plant traits. *Nature* **334**, 242–243.

Gaudet, C. L. & Keddy, P. A. (1995). Competitive performance and species distribution in shoreline plant communities: a comparative approach. *Ecology* **76**, 280–291.

Givnish, T. J. (1982). On the adaptive significance of leaf height in forest herbs. *American Naturalist* **120**, 353–381.

Goldberg, D. E. (1987*a*). Neighbourhood competition in an old-field plant community. *Ecology* **68**, 1211–1223.

Goldberg, D. E. (1987*b*). Seedling colonization of experimental gaps in two old-field communities. *Bulletin of the Torrey Botanical Club* **114**, 139–148.

Goldberg, D. E. (1990). Components of resource competition in plant communities. In *Perspectives on plant competition* (ed. J. Grace & D. Tilman), pp. 27–49. San Diego: Academic Press.

Goldberg, D. E. (1996). Simplifying the study of competition at the individual plant level: the consequences of distinguishing between competitive effect and response for forest vegetation management. *New Zealand Journal of Forestry Science* (In the press.)

Goldberg, D. E. & Barton, A. M. (1992). Patterns and consequences of inter-specific competition in natural communities: a review of field experiments with plants. *American Naturalist* **139**, 771–801.

Goldberg, D. E. & Fleetwood, L. (1987). Competitive effect and response in four annual plants. *Journal of Ecology* **75**, 1131–1143.

Goldberg, D. E. & Landa, K. (1991). Competitive effect and response: hierarchies and correlated traits in the early stages of competition. *Journal of Ecology* **79**, 1013–1030.

Goldberg, D. E. & Scheiner, S. M. (1993). ANOVA and ANCOVA: field competition experiments. In *Design and analysis of ecological experiments* (ed. S. Scheiner & J. Gurevitch), pp. 69–93. New York: Chapman & Hall.

Goldberg, D. E., Turkington, R. & Olsvig-Whittaker, L. (1995). Quantifying the community-level consequences of competition. *Folia geobotanica et phytotaxonomica* **30**, 231–242.

Goldberg, D. E. & Werner, P. A. (1983). Equivalence of competitors in plant communities: A null hypothesis and a field experimental approach. *American Journal of Botany* **70**, 1098–1104.

Grace, J. B. (1990). On the relationship between plant traits and competitive ability. In *Perspectives on plant competition* (ed. J. B. Grace & D. Tilman), pp. 51–65. San Diego: Academic Press.

Grace, J. B. (1995). On the measurement of plant competition intensity. *Ecology* **76**, 305–308.

Grace, J. B., Keogh, J. & Guntenspergen, G. R. (1992). Size bias in traditional analyses of substitutive competition experiments. *Oecologia* **90**, 429–434.

Grime, J. P. (1977). Evidence for the existence of three primary strategies in plants and its relevance to ecological theory. *American Naturalist* **111**, 1169–1194.

Grime, J. P. (1979). *Plant strategies and vegetation processes*. Chichester: John Wiley & Sons.

Gross, K. L. (1984). Effects of seed and growth form on seedling establishment of six monocarpic perennial plants. *Journal of Ecology* **72**, 369–388.

Gurevitch, J. (1986). Competition and the local distribution of the grass *Stipa neomexicana*. *Ecology* **67**, 46–57.

Gurevitch, J., Wilson, P., Stone, J. L., Teese, P. & Stoutenburgh, R. J. (1990). Competition among old-field perennials at different levels of soil fertility and available space. *Journal of Ecology* **78**, 727–744.

Herben, T. & Krahulec, F. (1990). Competitive hierarchies, reversals of rank order and the de Wit approach: are they compatible? *Oikos* **58**, 254–256.

Heske, E. J., Brown, J. H. & Mistry, S. (1994). Long-term experimental study of a Chihuahuan desert rodent community: 13 years of competition. *Ecology* **75**, 438–445.

Hubbell, S. P. & Foster, R. B. (1986). Biology, chance, and history and the structure of tropical rain forest tree communities. In *Community Ecology* (ed. T. Case & J. Diamond), pp. 314–329. New York: Harper and Row.

Inouye, R. S. & Tilman, D. (1995). Convergence and divergence of old-field vegetation after 11 yr of nitrogen added. *Ecology* **76**, 1872–1887.

Jacquard, P. (1968). Manifestation et nature des relations sociales chez les végétaux supérieurs. *Oecol. Plant.* **3**, 137–168.

Johansson, M. E. & Keddy, P. A. (1991). Intensity and asymmetry of competition between plant pairs of different degrees of similarity: an experimental study on two guilds of wetland plants. *Oikos* **60**, 27–34.

Keddy, P. A. & Shipley, B. (1989). Competitive hierarchies in herbaceous plant communities. *Oikos* **54**, 234–241.

Keddy, P. A., Twolan-Strutt, L. & Wisheu, I. C. (1994). Competitive effect and response rankings in 20 wetland plants: are they consistent across three environments. *Journal of Ecology* **82**, 635–643.

Law, R. & Watkinson, A. R. (1987). Response-surface analysis of two-species competition: an experiment on *Phleum arenarium* and *Vulpia fasciculata*. *Journal of Ecology* **75**, 871–886.

MacArthur, R. H. (1972). *Geographical ecology: Patterns in the distribution of species*. New York: Harper & Row.

Mahdi, A., Law, R. & Willis, A. J. (1989). Large niche overlaps among coexisting plant species in a limestone grassland community. *Journal of Ecology* **77**, 386–400.

Mahmoud, A. & Grime, J. P. (1976). An analysis of competitive ability in three perennial grasses. *New Phytologist* **77**, 431–435.

Marino, P. C. (1991). Competition between mosses (Splachnaceae) in patchy habitats. *Journal of Ecology* **79**, 1031–1046.

McConnaughay, K. D. M. & Bazzaz, F. A. (1990). Interactions among colonizing annuals: is there an effect of gap size? *Ecology* **71**, 1941–1951.

McGraw, J. B. & Chapin, S. F. I. (1989). Competitive ability and adaptation to fertile and infertile soils in two Eriophorum species. *Ecology* **70**, 736–749.

Mehrhoff, L. A. & Turkington, R. (1990). Microevolution and site-specific outcomes of competition among pasture plants. *Journal of Ecology* **78**, 745–756.

Menchaca, L. & Connolly, J. (1990). Species interference in white clover-ryegrass mixtures. *Journal of Ecology* **78**, 223–232.

Miller, T. E. & Werner, P. A. (1987). Competitive effects and responses between plant species in a first-year old-field community. *Ecology* **68**, 1201–1210.

Milne, A. (1961). Definition of competition among animals. In *Mechanisms in biological competition* (ed. F. L. Milthorpe), pp. 40–61. Cambridge University Press.

Pantastico-Caldas, M. & Venable, D. L. (1993). Competition in two species of desert annuals along a topographic gradient. *Ecology* **74**, 2192–2203.

Peart, D. R. (1989a). Species interactions in a successional grassland. II. Colonization of vegetated sites. *Journal of Ecology* **77**, 252–266.

Peart, D. R. (1989b). Species interactions in a successional grassland. III. Effects of canopy gaps, gopher mounds and grazing on colonization. *Journal of Ecology* **77**, 267–289.

Popma, J. & Bongers, F. (1988). The effect of canopy gaps on growth and morphology of seedlings of rain forest species. *Oecologia* **75**, 625–632.

Reader, R. J. (1993). Control of seedling emergence by ground cover and seed predation in relation to seed size for some old-field species. *Journal of Ecology* **81**, 169–175.

Rees, M. & Brown, V. K. (1992). Interactions between vertebrate herbivores and plant competition. *Journal of Ecology* **80**, 353–360.

Rice, K. J. & Menke, J. W. (1985). Competitive reversals and environment-dependent resource partitioning in Erodium. *Oecologia* **67**, 430–434.

Scandrett, E. & Gimingham, C. H. (1989). Experimental investigation of bryophyte interactions on a dry heathland. *Journal of Ecology* **77**, 838–852.

Shainsky, L. J. & Radosevich, S. R. (1992). Mechanisms of competition between Douglas-fir and red alder seedlings. *Ecology* **73**, 30–45.

Shipley, B. (1993). A null model for competitive hierarchies in competition matrices. *Ecology* **74**, 1693–1699.

Shipley, B. & Keddy, P. A. (1994). Evaluating the evidence for competitive hierarchies in plant communities. *Oikos* **69**, 340–345.

Shipley, B., Keddy, P. A. & Lefkovitch, L. P. (1991). Mechanisms producing plant zonation along a water depth gradient: a comparison with the exposure gradient. *Canadian Journal of Botany* **69**, 1420–1424.

Shmida, A. & Ellner, S. (1984). Coexistence of plant species with similar niches. *Vegetatio* **58**, 29–55.

Silvertown, J. & Dale, P. (1991). Competitive hierarchies and the structure of herbaceous plant communities. *Oikos* **61**, 441–444.

Smith, T. & Huston, M. (1989). A theory of the spatial and temporal dynamics of plant communities. *Vegetatio* **83**, 49–69.

Thompson, K. (1987). The resource ratio hypothesis and the meaning of competition. *Functional Ecology* **1**, 297–315.

Thompson, K. & Grime, J. P. (1987). Competition reconsidered – a reply to Tilman. *Functional Ecology* **2**, 114–116.

Tilman, D. (1987). On the meaning of competition and the mechanisms of competitive superiority. *Functional Ecology* **1**, 304–315.

Tilman, D. (1988). *Plant strategies and the dynamics and structure of plant communities*. Princeton University Press.

Tilman, D. (1990a). Mechanisms of plant competition for nutrients: the elements of a predictive theory of competition. In *Perspectives on plant competition* (ed. J. B. Grace & D. Tilman), pp. 117–141. San Diego: Academic Press.

Tilman, D. (1990b). Constraints and tradeoffs: toward a predictive theory of competition and succession. *Oikos* **58**, 3–15.

Tilman, D. & Wedin, D. (1991a). Dynamics of nitrogen competition between successional grasses. *Ecology* **72**, 1038–1049.

Tilman, D. & Wedin, D. A. (1991b). Plant traits and resource reduction for five grasses growing on a nitrogen gradient. *Ecology* **72**, 685–700.

Turkington, R. & Harper, J. L. (1979). The growth, distribution and neighbour relationships of *Trifolium repens* in a permanent pasture. IV. Fine scale biotic differentiation. *Journal of Ecology* **67**, 245–254.

Wedin, D. & Tilman, D. (1993). Competition among grasses along a nitrogen gradient: initial conditions and mechanisms of competition. *Ecological Monographs* **63**, 199–229.

Welbank, P. J. (1963). A comparison of competitive effects of some common weed species. *Annals of Applied Biology* **51**, 107–125.

Wilson, S. D. (1993a). Belowground competition in forest and prairie. *Oikos* **68**, 146–150.

Wilson, S. D. (1993b). Competition and resource availability in heath and grassland in the Snowy Mountains of Australia. *Journal of Ecology* **81**, 445–451.

Wilson, S. D. & Shay, J. M. (1990). Competition, fire, and nutrients in a mixed-grass prarie. *Ecology* **71**, 1959–1967.

Wilson, S. D. & Tilman, D. (1991). Components of plant competition along an experimental gradient of nitrogen availability. *Ecology* **72**, 1050–1065.

Wilson, S. D. & Tilman, D. (1995). Competitive responses of eight old-field plant species in four environments. *Ecology* **76**, 1159–1180.

Appendix I

List of studies included in the data base and summarized in Table 15.2. Variable definitions are given in Appendix II. (An asterisk before a reference indicates the study is part of the quantitative database (see text).)

Reference	Loc	Exp	#n	#t	Form	Even	Var	Time	Environ	Type	Transitiv. Eff	Transitiv. Res	Environ. Eff	Environ. Res	Traits
Aarssen 1988	g	add	4	4	herb	S–S	g	seas	–	exp	y	y	–	–	–
*Austin & Austin 1980	g	add	3	13	grass	S–S	g	seas	nutrient	exp	n	–	n	n	–
*Berkowitz et al. 1995	f	rem	1	3	h,s/t	A–S	gs	2.00	veg,time	nat	–	n	y?,n	–	–
*Bertness & Yeh 1994	f	rem	1	2	h,s	A–S	es	seas	salinity	nat	–	–	y?	–	–
*Bertness & Yeh 1994	f	rem	2	1	h,s/s	S–S	g	seas	water	exp	–	–	n	n	–
*Campbell & Grime 1992	cg	add	1	7	grass	S–S	gp	1.50	nutr,dist	exp	–	–	y	–	–
*Denslow et al. 1990	f	phy	1	7	tree	S–S	g	seas	–	–	–	–	–	–	rgr,sdwt
*de Steven 1991a,b	f	rem	1	6	h/t	A–S	egs	3.00	time	nat	–	–	n?	–	–
*Fowler 1990	f	add	4	4	grass	A–A	g	2.00	vegtype	nat	–	–	–	y	–
Gaudet & Keddy 1988[a]	g	add	10	2	herb	A–A	g	seas	–	–	–	y	–	–	mass,ht
*Gaudet & Keddy 1995	op	add	10	2	herb	A–S	g	seas	–	–	–	y	–	–	–
*Goldberg 1987a	f	rem	7	1	herb	A–A	g	seas	–	–	–	–	–	–	mass
Goldberg 1987b	f	rem	1	9	herb	A–S	s	seas	–	–	–	–	–	–	sdwt
*Goldberg & Fleetwood 1987	g	add	4	3	ann	S–S	g	seas	–	–	y	y	–	–	–
*Goldberg & Landa 1991	g	add	7	7	herb	S–S	g	seas	–	–	y	y	–	–	rgr,sdwt

Reference															sdwt
*Gross 1984	g	add	1	6	herb	A–S	g	seas	–	–	–	–	–	–	–	–
*Gurevitch 1986	f	rem	1	2	grass	A–A	g	3.00	water	nat	–	–	y	–	–	–
*Gurevitch et al. 1990	g	add	3	3	herb	S–S	g	seas	nutr,pot	exp	yn	yn	n?	y?	–	–
Johansson & Keddy 1991	g	add	6	6	herb	S–S	g	seas	–	–	y	y	–	–	–	–
*Keddy et al. 1994	g	add	20	3	herb	S–S	g	seas	nutr,dep	exp	yn	n	n	yn	–	–
*Law & Watkinson 1987	g	ser	2	2	ann	S–S	r	seas	–	–	y	y	–	–	–	–
Mahmoud & Grime 1976	g	add	3	3	grass	S–S	g	1.00	nutrient	exp	y	y	y	y	–	–
*Marino 1991	f	add	4	4	moss	S–S	gp	seas	water	exp	–	n	–	n?	–	–
*Marino 1991	g	add	4	4	moss	S–S	gp	1.50	water	exp	y	y	n	n	–	–
*McConnaughay & Bazzaz 1990	f	add	3	3	ann	S–S	gr	seas	gapsize	exp	y/	y	y?	y	–	–
*McGraw & Chapin 1989	f	rem	1	2	herb	A–A	gs	2.00	nutrient	nat	–	–	n	–	–	–
Mehrhoff & Turkington 1990	g	ser	2	2	herb	A–A	g	1.00	vegtype	nat	–	–	–	n	–	–
*Menchaca & Connolly 1990	g	ser	2	2	herb	S–S	g	seas	pH	exp	–	n	–	–	–	–
*Miller & Werner 1987	f	add	5	5	ann	S–S	g	seas	–	–	y	y	–	–	–	–
*Pantastico-Caldas & Venable 1995	f	add	2	2	ann	S–S	r	seas	water	nat	–	–	n?	y	–	–
*Peart 1989a	f	phy	4	4	grass	A–S	s	1.00	–	–	y	y	–	–	–	–
*Peart 1989b	f	rem	4	5	grass	A–S	s	1.00	–	–	–	n	–	–	–	–

Appendix I (cont.)

Reference	Loc	Exp	#n	#t	Form	Even	Var	Time	Environ	Type	Transitiv.		Environ.		
											Eff	Res	E ff	Res	Traits
Popma & Bongers 1988	f	phy	1	10	tree	A–S	g	seas	gapsize	nat	–	–	n	–	sdwt,ht
*Reader 1993	f	rem	1	12	herb	A–S	e	seas	herbivory	exp	–	–	n	–	sdwt
*Rees & Brown 1992	cg	rem	1	4	ann	S–S	gr	seas	herbivory	exp	–	–	y	n	–
Rice & Menke 1985	cg	ser	2	2	ann	S–S	g	seas	water	exp	–	n	–	n	–
*Scandrett & Gimingham 1989	g	ser	3	3	moss	S–S	gp	1.50	herb,light	exp	–	–	–	n	–
*Shainsky & Radosevich 1992	cg	ser	2	2	t,s	S–S	g	1.50	–	–	y?	y?	–	–	–
Shipley et al. 1991	f	rem	1	3	herb	A–A	g	1.25	wdepth	nat	–	n	y	–	–
*Turkington & Harper 1979	f	rem	4	4	herb	A–A	g	1.00	–	–	y	n	–	–	–
Welbank 1963	g	add	4	2	ann	S–S	g	seas	nutrient	exp	–	y	–	y	–
Wilson 1993a	f	rem	1	2	t,g	A–S	gs	seas	vegtype	nat	–	–	y	–	–
*Wilson 1993b	f	rem	1	3	t,g	A–S	g	1.50	nutrient	nat	–	–	y	–	–
*Wilson & Shay 1990	f	rem	1	2	herb	A–A	g	seas	nutrient	exp	–	–	y?	–	–
*Wilson & Tilman 1991	f	rem	1	3	herb	A–S	g	seas	nutrient	exp	–	–	y?	–	–
*Wilson & Tilman 1995	f	rem	1	8	herb	A–S	g	seas	nutr,dist	exp	–	–	n,y	–	–

a For testing correlations with traits, 44 neighbour species were used. Other traits that were correlated with competitive ability include shoot to root ratio, canopy diameter and area, leaf shape and size.

Appendix II

Definitions and criteria for variables in Appendix I. Only options that are abbreviated in Appendix I and possibly ambiguous are noted here.

Reference: * The study is included in the quantitative data base (published in *Ecology* or *Journal of Ecology* between 1979 and 1995).

Loc (location): f, field; g, glasshouse; cg, common garden; op, pots placed outdoors.

Exp (type of experiment): add, additive (targets at constant density and neighbour density manipulated by planting, adding individuals); rem, removal (targets at constant density and neighbour density manipulated by removal of existing plants; phy, phytometer (targets introduced at a constant density but neighbour density differs due to disturbances, e.g. treefall gaps versus undisturbed forest, gopher mounds versus undisturbed grassland); ser, addition series (both density and frequency of targets and neighbours experimentally manipulated).

#n (number of neighbour taxa): taxa are usually species but may be genotypes or groups of species (e.g. #n = 1 for a field experiment usually indicates all vegetation was present versus removed).

#t (number of target taxa): taxa are usually species but may be genotypes or groups of species.

Form (life/growth form of neighbour and target taxa): ann, annual; grass or g, graminoids; herb or h, herbaceous (including graminoids + dicots); s, shrub; t, tree. Where growth forms are separated by a slash, the value before the slash is the neighbour growth form and the value after the slash is the target growth form.

Even (evenness of initial size/stages of interacting taxa): A–A, adult–adult (includes target transplants of clonal plants that might be initially smaller than surrounding vegetation; A–S, adult neighbours and seed or seedling targets; S–S, seed or seedlings interacting with other seeds or seedlings.

Var (dependent variable): e, emergence; g, growth or size; s, survival; r, reproductive output; gp, 'population' growth or abundance (data reported per unit area rather than per plant).

Time (duration of the experiment): number of years; seas, one growing season or less.

Environ (environmental differences if competitive abilities were measured in more than one environment): nutr, nutrient; wdepth or dep, water depth; dist, disturbance; veg or vegtype, type of vegetation (usually confounded with other environmental differences); time, experiment repeated at different times; pot, pot size.

Type (origin of environmental differences): exp, experimentally manipulated; nat, naturally occurring.

Transitivity-Eff (consistency of competitive effects of neighbours for different targets): y, consistent; n, contingent; yn or ny, variable results among independent tests (e.g. for different environments). ? modifying y or n indicates that the conclusion is by inspection of data and not confirmed statistically.

Transitivity-Res (consistency of competitive responses of targets for different neighbours): as for Transitivity-Eff.

Environment-Eff (consistency of competitive effects among environments): as for Transitivity-Eff. Values separated by a comma are for different kinds of environmental treatments with results following the same order as the listing of environmental treatments under 'Environ'.

Environment-Res (consistency of competitive responses among environments): as for Environment-Eff.

Traits (traits quantitatively related to competitive ability): sdwt, seed mass; ht, height; mass, mass per plant; rgr, relative growth rate. The last three are usually measured in the absence of competition.

Index

α, sexual maturity age, defined, 215
Acer spp., phylogenetic tree (*illus.*), 24–5, 27
actinorhizal associations, *Alnus*, 272
Aegopodium podagraria, clonal structure
   (*illus.*), 195
alien floras, British Isles, comparative
   ecology, 36–53
alien species, defined, 37
alkaloids, data for apparency hypothesis,
   9–12
allele frequencies, 244
alleles
   aplastic optimal, 234
   fitness landscapes (*illus.*), 235
   optimal, 231
allelic fitness, 230, 243
   optimal plastic strategy, reproductive
     value, 249–50
allometry of life history trade-offs, 176–8,
   179
allozyme diversity, 103
   variation among populations (*illus.*),
     92–3
Alnus, actinorhizal, 272
angiosperms
   analysis of orders, 274
   distribution of clonality (*illus.*), 199
   first, magnoliid vs palaeoherb, 23–4
   phylogenetic trees
   mycorrhizas (*illus.*), 277
   non-mycorrhizal species, 277–9
aphids, melaphidine, 256
'apparency' hypothesis, herbivory, 9–10,
   253
Aquilegia, 74
archaeophytes, defined, 37
arctic habitats, adult persistence, 192, 193
arthropods, herbivory, 253

Asclepias, ESS, inflorescence size, 182–6
Asperula odorata, clonal structure (*illus.*),
   195
Asteraceae
   annual/perennial contrasts, 8–9
   genetic variation (*table*), 116
   phylogeny, 66
   (*illus.*), 7
   phylogenetic inertia, 61
autogamy *see* self-pollination

background selection, hitchhiking and
   genetic drift (*table*), 91
breeding system
   and geographic range, 110, 111
   life form, 111
   and seed dispersal (*table*), 107, 108
   and taxonomic status, 108, 109, 110
British Isles, comparative ecology of native
   and alien floras, 36–53
   aliens
   controlled contrasts, life history traits
     (*table*), 47
   distribution (*illus.*), 45
   data sources, 37–9
   geographic origin of aliens, 42
   habitats invaded (*illus.*), 42–3
   introductions
     modes, 41–2
     numbers, 40, 43–4
   pest species, 39
   Raunkaier life forms (*illus.*), 41, 46, 48

CAIC package, (comparative analyses of
   independent contrasts), 126–7, 197,
   201
Caltha palustris, clonal structure (*illus.*),
   195